Elemental Future

Navigating the Resources of Tomorrow

By
Henry O'Neal

Elemental Future

Navigating the Resources of
Tomorrow

Table of Contents

Introduction

As we stand at the precipice of the 21st century's most profound challenges and opportunities, understanding the future of technology and sustainability becomes paramount. We find ourselves grappling with a paradox: how can we continue to drive technological progress while ensuring the health and longevity of our planet? This book delves into this intricate and often daunting question, aiming to illuminate the path forward.

The essence of our technological advancements lies in the raw materials that fuel innovation. From rare earth elements to emerging bioplastics, these materials are the building blocks of our future. But with this dependency comes a responsibility—one that transcends industries and borders, touching every individual who benefits from modern technology. We must consider not just the immediate utility of these materials, but their long-term impacts on our environment, our economies, and our societies.

Our world is in the midst of a remarkable transformation. The digital revolution has altered how we live, work, and interact, while concurrently, the imperative for sustainability has never been more urgent. The juxtaposition of rapid technological advancements with the pressing need for environmental stewardship is at the heart of this book. By exploring the critical materials essential for tomorrow's technologies, we aim to equip you with the knowledge to make informed decisions and inspire actions towards a sustainable future.

Consider the smartphone in your hand or the electric vehicle silently cruising through your neighborhood. These marvels of modern engineering are just the surface level of a complex web of resources, logistics, and innovation. Every battery powering these devices, every circuit board ensuring their functionality, and every solar panel harnessing the power of the sun relies on specific materials—often rare, sometimes environmentally damaging, and invariably essential. Understanding these materials and the implications of their extraction, use, and disposal is crucial.

The rapid pace of technological evolution means that what is cutting-edge today might be obsolete tomorrow. However, the materials that underpin these technologies will continue to play a crucial role. As such, a fundamental aspect of our discussion will revolve around sustainability. How can we ensure that our pursuit of the next big thing does not compromise our planet's health? This book will explore innovative approaches to recycling, recovery, and the development of alternative materials that promise to reduce our ecological footprint.

Consider the solar revolution. Solar panels now adorn rooftops across the globe, capturing the sun's energy and converting it into clean electricity. Yet, the story of solar technology is not just about capturing sunlight. It's a tale of material science, where the quest for more efficient and durable materials like perovskites is continuously pushing the boundaries. Similarly, wind energy, with its towering turbines, relies heavily on composite materials to optimize performance and longevity. These intersections of material science and sustainable energy are pivotal to our future.

Water, too, is a critical resource in the narrative of sustainable technology. From hydroelectric dams to innovative desalination plants, the ways we harness and utilize water for energy speak volumes about our progress and our potential pitfalls. Understanding the nexus between water and energy provides insights into opportunities for innovation and efficiency that can drive us toward a more sustainable future.

In addressing these multifaceted topics, we will also consider the role of policy and governance. Sustainable management of resources requires more than just technological innovation; it demands a framework of policies that foster cooperation and responsible use. Through the lens of international cooperation and strategic policy-making, we will explore how nations can collaborate to address resource scarcity and propel innovation.

Equally important is the human element—the engineers, the policymakers, and the everyday citizens whose decisions collectively shape our world. Preparing the workforce of tomorrow involves not just technical skills but an ethical understanding of resource management. Engaging communities and individuals in sustainable practices will be key to effectuating lasting change. Together, these elements form the bedrock of a future where technology and sustainability are not at odds but are seamlessly integrated.

Moreover, the circular economy presents a paradigm shift from the traditional linear model of 'take, make, dispose' to one where resources are kept in use for as long as possible. By redesigning products for durability and ease of recycling, we can minimize waste and create a closed-loop system that promotes resource efficiency. This book will delve into how innovative recycling technologies and principles of circular economies can herald a new era of sustainable resource management.

The economic dimensions of sustainable materials cannot be overlooked. Investments in new materials and sustainable technologies offer not only environmental benefits but also significant economic opportunities. By focusing on sustainable business models and funding mechanisms, we can catalyze the development of technologies that are both economically viable and ecologically sound.

As we explore these themes, one thing becomes clear: the interconnectedness of global resources. Decisions made in one part of the world can have far-reaching consequences, underscoring the need for a global perspective on resource management. Ensuring equitable access to resources and fostering international collaboration will be critical in addressing the challenges ahead.

This book is not just a compendium of technological and scientific knowledge; it's a call to action. We hope to inspire you to think critically about the materials that shape our world and the sustainable practices that can secure our future. By understanding the intricacies of these resources and their applications, we can collectively drive the shift towards a more sustainable, innovative, and environmentally conscious society.

In the coming chapters, we will peel back the layers on everything from the geopolitics of rare earth elements to groundbreaking advancements in nuclear fusion. Each chapter is designed to provide you with a thorough understanding of the critical materials driving technology and the sustainable practices that can enhance their use. Our aim is to inform, enlighten, and inspire you to play a part in building a future where technological progress and environmental stewardship go hand in hand.

The journey towards sustainability is complex and multifaceted, but it is also filled with hope and potential. Together, let's discover the raw materials of the future and explore how we can harness them responsibly for the betterment of our world. The stakes are high, but so too are the possibilities. Welcome to the intersection of technology and sustainability, where the future is being forged today.

Chapter 1:
Unveiling the Future of Resources

As we stand on the precipice of a new era in technological advancement, the stakes have never been higher for the sustainable management of our planet's vital resources. Modern technology demands an array of critical materials, from rare earth elements to advanced composites, shaping devices and systems that drive our daily lives. Yet, the pursuit of technological innovation must be balanced with environmental stewardship. The future isn't just about the gadgets we create, but also about how we source, use, and reuse the materials that underpin them. This delicate balance is the crux of designing a future that's both innovative and sustainable. By exploring the intricate interdependencies of modern resources and their impact on our ecological footprints, we can better appreciate the challenges and opportunities ahead. Our global community must collaborate to redefine resource efficiency, transforming scarcity into abundance through novel strategies and technologies. This chapter initiates a journey to uncover the materials that will not only propel the next technological wave but also sustain our world for generations to come.

the role of critical materials in modern technology

The progression of modern technology hinges on an intricate web of materials, each playing a unique and indispensable role. From the smartphones in our pockets to the advanced medical devices that save lives, critical materials underpin the innovations we often take for granted. These materials, frequently sourced from rare earth elements, lithium, and other less common minerals, are pivotal to the functionality, efficiency, and advancement of contemporary technology.

Rare earth elements (REEs) are a prime example of critical materials with profound implications for modern technology. Comprising 17 chemical elements, REEs are integral to manufacturing high-strength magnets, catalysts, and phosphors used in a variety of applications, including electric vehicles, wind turbines, and LED lighting. Their unique properties enhance the performance and efficiency of these technologies, making them essential for the green energy transition.

Imagine your daily routine: waking up to an alarm on a smartphone, brewing a cup of coffee with an electric coffee maker, commuting in an electric vehicle, and working on a laptop. At each step, critical materials are at work. The smartphone's screen and battery rely on indium and lithium, the coffee maker's heating elements might use nickel or copper, and electric vehicles are powered by lithium-ion batteries. Without these materials, our modern conveniences would be far less advanced or even nonexistent.

Silicon is another cornerstone of modern technology. As the primary component in semiconductors, silicon drives the digital world, powering everything from computers and smartphones to the servers that support the internet. While silicon itself is not rare, the demand for high-purity silicon has escalated with the proliferation of digital devices. The ongoing push to smaller, more efficient chips further underscores the importance of this material in maintaining and advancing technological growth.

Beyond consumer electronics, critical materials play a vital role in healthcare. For example, titanium is widely used in medical implants due to its biocompatibility and strength. MRI machines rely on superconducting magnets made from niobium-titanium alloys. The development of advanced imaging and diagnostic tools enhances our ability to detect and treat diseases, improving patient outcomes and reducing healthcare costs.

Renewable energy technologies, central to combating climate change, also depend heavily on critical materials. Wind turbines require neodymium and dysprosium for their powerful magnets. Solar panels utilize indium, gallium, and selenium to convert sunlight into electricity efficiently. These materials are not just supporting technologies; they are revolutionizing the way we produce and consume energy, steering us towards a more sustainable future.

However, the reliance on critical materials comes with significant challenges, notably supply chain vulnerabilities and environmental impacts. Many of these materials are concentrated in specific geographic locations, making their supply susceptible to geopolitical tensions and trade conflicts. The extraction and processing of these materials often entail substantial environmental and social costs, such as habitat destruction, pollution, and labor exploitation.

Addressing these challenges necessitates a multifaceted approach. Diversifying the supply chain by developing alternative sources, investing in recycling and recovery technologies, and promoting sustainable mining practices are crucial steps. Additionally, advancing research in material science to find substitutes or develop new materials with similar or superior properties can mitigate the reliance on specific critical materials.

For instance, researchers are exploring ways to recycle rare earth elements from electronic waste, reducing our dependency on freshly mined materials. Innovations in battery technology, such as solid-state batteries, aim to use more abundant and environmentally friendly materials. Progress in these areas not only helps secure the supply of critical materials but also aligns with broader sustainability goals, reducing the ecological footprint of modern technology.

Education and awareness are equally important in tackling the challenges associated with critical materials. By understanding the complex journey from raw material to finished product, consumers and policymakers can make informed decisions that support sustainable practices. This knowledge can drive demand for products designed with longevity, recyclability, and minimal environmental impact in mind.

Governments and industries must collaborate to create robust policies and standards that ensure the ethical and sustainable sourcing of critical materials. This includes enforcing regulations that protect the environment and worker rights, incentivizing recycling programs, and fostering innovation in material science and engineering. Strategic stockpiling and international cooperation can also buffer against supply disruptions and ensure a steady flow of essential materials.

The role of critical materials in modern technology is undeniably crucial, yet it highlights a delicate balance. We must harness these materials' potential to drive innovation and sustainability and be mindful of their environmental and social impacts. By investing in research, education, and policy development, we can pave the way for a future where technology thrives without compromising the planet's health and resources.

In conclusion, critical materials are the lifeblood of modern technology, indispensable for the continued advancement of various sectors, from consumer electronics and healthcare to renewable energy. The sustainable management of these resources will determine our ability to innovate responsibly and achieve a balance between technological progress and environmental stewardship. As we look forward, the choices we make today regarding the use and preservation of critical materials will shape the technological landscape of tomorrow.

Innovation and sustainability are not mutually exclusive; they must go hand in hand to ensure a resilient and prosperous future. It is an exciting time for material science, full of challenges and opportunities to redefine what is possible through conscientious and creative thinking. By leaning into this collaborative and forward-thinking ethos, we can unlock the true potential of critical materials, spearheading a technological revolution that respects and safeguards our planet.

The responsibility lies with all stakeholders—researchers, policymakers, industries, and consumers—to drive the change towards sustainable resource management. Together, we can transform the narrative from exploitation to innovation, from consumption to regeneration, and from scarcity to abundance. The future of resources, and by extension, the future of technology, depends on the actions we take today.

Exploring the Frontiers of Sustainability

As we traverse the uncharted territories of the future, sustainability emerges as a non-negotiable compass guiding our way. It's clear that the resources we rely on today have profound implications for the environment, economy, and society. Thus, understanding how we can sustainably manage and utilize these resources is vital if we are to build a resilient future.

What does it mean to explore the frontiers of sustainability in the context of future resources? It's about looking beyond traditional practices and ideating new ways of doing things that minimize our ecological footprint. This exploration involves integrating advanced technologies, innovative materials, and sustainable practices into every aspect of resource utilization.

One of the striking elements at the frontier of sustainability is the shift from linear to circular economies. Instead of the traditional 'take, make, dispose' model, circular economies advocate for designing waste out of the system. The emphasis is on reusing, refurbishing, and recycling materials, extending the life cycle of products and reducing the demand for virgin resources.

In parallel, there's a growing push towards harnessing renewable energy sources. Solar and wind energy technologies have seen remarkable advancements, but the journey doesn't stop there. We're constantly pushing the boundaries with next-generation materials like perovskites in solar cells and advanced composites in wind turbines. These innovations are not just about efficiency but also about creating systems that have minimal environmental impact over their lifetime.

Sustainable mining practices also play a crucial role in this exploration. Traditional mining is resource-intensive and environmentally damaging. However, green mining technologies are emerging, which focus on reducing environmental impact through innovative methods. Urban mining, for instance, presents an exciting opportunity by unlocking materials found in electronic waste and other urban sources, thus diverting waste from landfills and recovering valuable resources.

On the horizon, bio-based and biodegradable materials offer a promising path forward. These materials, derived from renewable sources like plants, can significantly lower our dependence on fossil fuels. However, the challenge lies in scaling up production and ensuring that these materials can meet the demands of modern technology without compromising performance or longevity.

Moreover, the quest for sustainability touches upon the very water and air that sustain life. Advancements in water purification technologies and air purification materials are not only protecting these essential resources but also ensuring their availability and quality for future generations. This is critical in regions facing severe water stress and air pollution challenges.

Equally important is the role of education and ethical considerations in sustainability. Preparing the workforce of tomorrow requires not just technical training but also a deep understanding of the ethical implications of resource usage. Encouraging a culture of sustainability starts with education that fosters innovative thinking and a commitment to ethical responsibility.

To truly explore the frontiers of sustainability, international cooperation and robust policy frameworks are indispensable. Resource management is a global challenge that transcends borders. Policies that support sustainable practices and innovations can make a significant difference, particularly when nations come together to address resource scarcity and ecological degradation collectively.

Investment in sustainable technologies and green innovations is another cornerstone of this exploration. Funding the future means investing in disruptive technologies and supporting startups and researchers pushing the boundaries of what's possible. This not only accelerates the development of sustainable solutions but also creates economic opportunities and jobs in green sectors.

In the context of cities, smart materials and urban planning are reshaping how we think about urban landscapes. Integrating green spaces, developing efficient public transportation systems, and using smart materials in construction are all part of building cities that are resilient, resource-efficient, and liveable.

Transportation is yet another critical frontier. The transition to electric vehicles and sustainable aviation fuels encapsulates the ongoing shift towards more sustainable modes of travel. This transition requires advancements in battery technology, alternative fuels, and infrastructure development, ensuring that these solutions are accessible and practical.

Artificial intelligence (AI) and big data are proving to be invaluable allies in the journey towards sustainability. From optimizing resource use to enhancing recycling and waste management processes, AI is transforming how we approach some of the most pressing sustainability challenges.

The interconnectivity of global resources means that local decisions can have far-reaching impacts. Ensuring equitable access to resources and fostering global cooperation are essential to addressing the intertwined challenges of resource scarcity, environmental conservation, and economic development.

Despite the progress, significant challenges remain. Overcoming technical and economic barriers, and fostering societal adaptations to new paradigms of resource management require sustained effort, innovation, and collaboration. These challenges, however, also present opportunities to learn and improve.

By showcasing success stories and lessons learned from around the world, case studies offer valuable insights into what works and what doesn't. They highlight how communities, companies, and nations are pioneering sustainable practices and providing blueprints for others to emulate.

Imagining a sustainable world is not a distant dream but a feasible reality that we can achieve through collective effort. The steps we take today, guided by visionary thinking and grounded in pragmatic action, will determine the future of resources and the health of our planet.

In conclusion, the journey towards sustainability is an ongoing exploration. It requires bold thinking, innovative solutions, and unwavering commitment to creating a world where technology and nature harmoniously coexist. By embracing sustainability at the frontiers of resource management, we pave the way for a future that's not only technologically advanced but also ecologically secure and socially equitable.

Chapter 2:
The Pillars of Tomorrow's Technologies

As we delve deeper into the evolving landscape of modern advancements, it's apparent that the bedrock of future technologies rests heavily on key materials that will drive innovation. Rare earth elements, though scarcely mentioned in everyday dialogue, form the backbone of many high-tech devices, quietly enabling progress in realms we've grown to rely on. Equally transformative is lithium, catalyzing the battery revolution and propelling us toward an era of clean energy and electric mobility. These critical substances aren't merely raw materials; they are the silent architects of tomorrow's world, shaping everything from our energy infrastructure to the consumer gadgets we can't imagine living without. Truly understanding, managing, and innovating around these pillars is essential not just for technological advancement but for sustainable stewardship of our planet's finite resources. As we navigate the future, the intersection of cutting-edge technology and responsible resource management will define the legacy we leave for the generations to come.

Rare Earth Elements: The Unsung Heroes

Rare earth elements might not be household names, but they are indispensable to the advancements underpinning modern technology. These 17 chemical elements, nestled at the far end of the periodic table, are critical in manufacturing everything from smartphones to electric vehicles and wind turbines. Their unique magnetic, luminescent, and electrochemical properties make them the silent enablers in the quest for a more sustainable future. Despite their name, rare earth elements are relatively abundant in the Earth's crust, but their extraction is complicated and labor-intensive, leading to significant geopolitical and environmental challenges. As we forge ahead into tomorrow's technologies, developing efficient and environmentally friendly methods to recycle and recover these elements will be vital. Their pivotal role reminds us that the journey toward sustainability isn't just about harnessing new energy sources, but also about responsibly managing the materials that power our innovations.

The Geopolitics of Rare Earth Elements have positioned themselves as fundamental players in the tech-driven world we're evolving into. Their acquisition and control are not merely scientific endeavors but are heavily intertwined with international power dynamics and economic strategies.

In recent years, global dependence on rare earth elements (REEs) for advanced technologies has raised the stakes in geopolitical terms. These 17 elements, while scattered across the planet, are not evenly distributed. Countries rich in REEs possess a significant leverage in the technological and economic landscapes of the future.

The geopolitical struggle for these elements is predominantly exemplified by China's extensive control over REE mining and production. Currently, China provides more than 80% of the world's rare earth supply. This near-monopoly allows it to manipulate the global market, affecting prices and availability, which subsequently influences technological innovation worldwide. Case in point: when China restricted its REE exports in 2010, prices soared, impacting various tech industries and highlighting global vulnerabilities.

China's dominant position isn't just a product of geography but also of strategic national policies. For years, China has been investing heavily in the mining and processing of rare earths, establishing a robust infrastructure and nurturing a skilled workforce. They've created vast mining operations and processing plants that other countries have struggled to match, both in scale and efficiency. This strategic foresight places China at a critical junction of future technological advancements.

However, this concentrated control has provoked a global response. Nations like the United States, Japan, and members of the European Union are actively seeking to diversify their REE sources and reduce dependency on China. These efforts include reopening old mines, investing in REE extraction technologies, and forging alliances with other resource-rich countries. For instance, the United States has been reinvesting in domestic production capacities, with projects in California aiming to restore some degree of self-sufficiency.

Beyond the immediate players, countries with latent REE deposits could emerge as vital new sources in this geopolitical chess game. Nations such as Australia and Canada have untapped reserves that are becoming increasingly important. These countries, recognizing the critical nature of REEs, are investing in developing their mining sectors and positioning themselves as alternative suppliers.

Africa too is gaining attention. Rich in resources and historically underexplored in terms of REE potential, countries like South Africa and Tanzania are becoming significant points of interest. International mining companies are now eyeing these nations, bringing both opportunities and challenges. While this could spur economic growth and development, it also raises concerns about environmental sustainability and economic exploitation. Sustainable extraction practices must be prioritized to avoid repeating the ecological and social mistakes seen in other regions.

In addition to sourcing diversification, there is a burgeoning interest in recycling and recovering rare earth elements from electronic waste. This approach not only addresses the supply chain vulnerability but also promotes sustainable practices. Japan has notably led this initiative, developing sophisticated recycling methods to reclaim REEs from discarded electronics.

Market dynamics and trade policies also play a critical role. Tariffs, trade agreements, and export controls can drastically reshape the REE landscape. For example, the ongoing trade tensions between the United States and China have underscored how geopolitical decisions influence global technology supply chains. Policies enhancing collaboration and open markets are essential, as are strategies to buffer against abrupt supply disruptions.

Aside from the supply concerns, the processing and refining of REEs present additional geopolitical sensitivities. These processes are complex, expensive, and often environmentally hazardous. Thus, countries that can refine REEs have an added layer of advantage. Current dominance by China in this sector forces other nations to either export mined REEs for processing, often back to China, or invest heavily in developing their own refining capabilities, which could take years and substantial financial resources.

Geopolitical complexities are further exacerbated by the strategic importance of REEs in defense technologies. From advanced missile systems to sophisticated radar and communication equipment, military applications rely on these elements. Controlling REE supply chains becomes not only an economic advantage but a national security imperative. Countries without stable access to these materials may find themselves at a strategic disadvantage.

This profound global interdependence on REEs demonstrates how interconnected technological progress and resource geopolitics are. The urgency to secure rare earth elements has led to a broader realization among nations: a collaborative approach may be necessary to establish a more sustainable and stable supply chain.

International cooperation could include joint ventures in mining, shared investments in refining technologies, and collective efforts in recycling and material substitution research. As nations begin to recognize the shared interest in securing supplies without depleting the planet, there is an opportunity to redefine resource geopolitics in a way that mitigates conflict and fosters shared technological advancement.

The geopolitical terrain for REEs is uncharted and continuously evolving. The interplay between national policies, global markets, and technological innovations reflects a broader narrative of 21st-century resource management. To navigate these waters effectively will require wisdom, foresight, and a concerted effort to balance industrial progress with sustainable practices.

Ultimately, the geopolitics of rare earth elements are shaping the future of technology and sustainability. By understanding and addressing these complex dynamics, we can create a world where innovation and resource management coexist, forging a path towards a truly sustainable global society.

Recycling and Recovery: A Sustainable Approach isn't just a phrase; it's a guiding principle for a future where resources are cherished and reused. In the realm of rare earth elements, this approach isn't just a moral choice; it's a practical necessity. Recycling these critical materials isn't merely about reducing waste; it's about creating a sustainable loop that minimizes dependency on finite natural deposits and reduces geopolitical tensions.

The demand for rare earth elements has surged, driven by burgeoning industries like electric vehicles, wind turbines, and advanced electronics. These materials are vital for the production of high-performance magnets, batteries, and other essential components.

However, as availability dwindles, the need to recycle and recover these elements becomes increasingly clear. By adopting a comprehensive recycling strategy, industries can mitigate supply shortages and foster a more resilient supply chain.

Technological advancements are making recycling processes more efficient and economically viable. For example, methods like hydrometallurgical and pyrometallurgical processes are being optimized to recover rare earth elements from discarded electronics and industrial waste. These processes can extract valuable materials while minimizing environmental impact, thus aligning economic benefits with ecological responsibility.

In addition, research is being conducted into biological methods for recycling rare earth elements. Certain bacteria have shown promise in leaching these elements from ores and waste, providing a potentially low-energy and low-impact alternative to traditional methods. The integration of biotechnological solutions could revolutionize recycling paradigms, making it possible to recover rare earth elements from sources that were previously deemed too challenging or uneconomical to process.

Recycling rare earth elements also aligns with the principles of a circular economy. In a circular model, products are designed for durability, repairability, and eventual recycling. This not only extends the life of products but also ensures that materials are continuously cycled through the economy. Designing electronics and other high-tech goods with end-of-life recovery in mind is a proactive step toward this goal.

Moreover, policy frameworks play a crucial role in promoting recycling and recovery. Governments and regulatory bodies can incentivize sustainable practices through subsidies, tax breaks, and strict environmental regulations. The creation of standardized recycling protocols and the enforcement of e-waste management laws can further drive the adoption of these practices worldwide.

Corporate responsibility is equally essential. Companies must recognize their role in the broader ecological context and adopt sustainable practices willingly. Many leading tech companies are already investing in closed-loop supply chains, where the goal is to reclaim as much material as possible from end-of-life products. This shift isn't just about compliance; it's about aligning business operations with the planet's long-term needs.

Recyclable materials don't just come from old electronics and industrial scrap; urban mining is another burgeoning field. Urban mining refers to reclaiming valuable materials from existing urban infrastructure, such as buildings, vehicles, and even landfills. By tapping into these 'above-ground mines', cities can recycle an array of important resources, contributing to urban sustainability and reducing the need for new extraction projects.

Public awareness and participation are also key to the success of recycling programs. Educational campaigns that inform consumers about the importance of recycling and how to properly dispose of e-waste can dramatically increase recycling rates. Community-based collection programs provide convenient ways for individuals to participate in the recycling loop, turning individual actions into collective impact.

While the technical aspects of recycling and recovery are vital, the economic aspects can't be overlooked. Initial investments in recycling technologies and infrastructure can be substantial, but the long-term economic benefits often outweigh these upfront costs. Recycled materials can considerably lower the raw material costs for manufacturers, while also reducing the environmental cleanup costs associated with traditional mining.

Global collaboration is another cornerstone of this sustainable approach. Rare earth element reserves are unevenly distributed across the globe, but the demand for these materials is nearly universal. International partnerships and agreements can help ensure that recycling technologies and recovered materials are shared equitably, benefiting all parties involved and fostering global stability.

Industry standards and certifications can help create a level playing field. Certifications such as 'recycled content' labels can help consumers make informed choices and encourage manufacturers to meet higher standards of sustainability. Transparency in supply chains ensures that recycled materials are used ethically and effectively.

Investments in research and development are essential to push the boundaries of what recycling can achieve. Continuous innovation is needed to refine existing methods, discover new ones, and scale up successful technologies. Governments, academia, and private sectors must collaborate to fund and drive this critical research.

Lastly, the move towards a sustainable recycling and recovery system is about making tough yet necessary choices. It's about choosing long-term gains over short-term profits and prioritizing ecological balance over unchecked industrial growth. This sustainable approach demands collective action, guided by foresight and responsibility.

As we stand on the cusp of revolutionary advancements in technology, our approach to resource management will define the success of these innovations. By embracing recycling and recovery, we not only preserve the materials needed for future technologies but ensure those technologies contribute to a world where sustainability isn't an afterthought but a fundamental principle.

Lithium and the Battery Revolution

The world is on the cusp of a transformation driven by lithium-ion batteries, underscoring lithium's pivotal role in future technologies. As our reliance on renewable energy intensifies, these batteries are not just powering electric vehicles; they're revolutionizing energy storage systems, ensuring that solar and wind energy can be harnessed efficiently. Lithium's lightweight property and high energy density make it ideal for this purpose. But with its growing demand comes responsibility: we must explore sustainable mining techniques, address environmental impacts, and investigate viable alternatives. By prioritizing the ethical sourcing and recycling of lithium, we can propel this battery revolution towards a greener, more resilient future, ultimately balancing technological advancement with environmental stewardship.

From Ore to Electrification: The Lithium Journey has the power to reshape the landscape of modern technology and energy usage, creating a revolutionary shift. The pathway from raw lithium ore to the active battery cells powering our latest gadgets and electric vehicles traverses a complex, multi-step process that integrates both advanced technology and environmental challenges. It's a journey worth understanding as we consider the sustainability and future impact of this incredible element.

Beginning with the mineral source, lithium predominantly comes from two types of deposits: hard rock and brine. Hard rock mining involves extracting lithium ore from spodumene, a mineral found in pegmatite rocks. These rocks are then processed to extract lithium hydroxide or lithium carbonate. On the other hand, brine extraction entails pumping mineral-rich water from underground deposits into evaporation ponds, where the water gradually evaporates, leaving behind concentrated lithium salts. Both methods present unique challenges and environmental concerns, including land disruption and water usage.

Once extracted, the raw lithium undergoes a series of chemical treatments to enhance its purity. For example, lithium carbonate or lithium hydroxide is extracted through careful refining processes. These compounds are critical in the production of lithium-ion batteries, which have become the cornerstone of modern energy storage solutions. The chemicals derived from lithium ore undergo stringent quality control measures to ensure their suitability for high-performance applications.

The refined lithium then makes its way into battery production facilities, marking the start of its transformation into a component of advanced energy storage systems. Lithium-ion batteries consist of several key elements, including the cathode, anode, electrolyte, and separator. Each of these components must be meticulously crafted and combined to form a battery cell. The cathode, typically made of lithium cobalt oxide, lithium iron phosphate, or other lithium compounds, plays a vital role in the battery's overall energy capacity and lifespan.

The assembly of lithium-ion batteries is a delicate and precise process. Battery manufacturers must achieve the perfect balance between material quality, electrode design, and manufacturing precision. Any deviations can lead to sub-optimal performance or, worse, safety concerns. This is why the industry continuously invests in research and development to improve battery efficiency, energy density, and safety standards.

Once manufactured, lithium-ion batteries enter a broad array of applications, each contributing to the electrification of different aspects of modern life. The most notable use is in electric vehicles (EVs), where the demand for efficient and long-lasting batteries is ever-increasing. Through rigorous testing and validation, these batteries are optimized to provide the best possible range and performance, pushing the boundaries of automotive technology.

However, the journey of lithium doesn't end with the battery's initial use. As these batteries wear out or become obsolete, recycling becomes paramount to ensure the sustainability of lithium resources. Recycling lithium-ion batteries involves retrieving valuable materials, including lithium, cobalt, nickel, and manganese. This process not only mitigates environmental impact but also reduces the need for fresh mineral extraction, fostering a circular economy.

The environmental implications of lithium extraction and battery production can't be overlooked. While lithium-ion batteries offer significant environmental advantages over fossil fuels, their production and disposal carry potential ecological risks. Innovations and regulations are crucial in mitigating these impacts. For instance, advances in green chemistry and waste management are making strides in reducing the environmental footprint of lithium battery production.

Exploring the potential alternatives to lithium-ion technology is another essential aspect of this journey. Researchers are investigating novel materials and electrolytes, such as solid-state batteries and lithium-sulfur cells, which promise higher energy densities and enhanced safety profiles. These innovations could potentially alleviate some of the environmental concerns associated with traditional lithium-ion batteries.

The entire lifecycle of lithium, from ore to electrification, underscores the need for a balanced and forward-thinking approach. Investment in research and development, adherence to stringent environmental standards, and fostering international cooperation are essential elements in ensuring lithium's sustainable future. Strategic partnerships between governments, industries, and research institutions can accelerate technological advancements and promote responsible resource management.

In this context, lithium stands as both a challenge and an opportunity. Its mining and processing demand careful consideration of environmental practices, yet its vital role in energy storage solutions paves the way for a greener future. Policymakers and industry leaders must pave the way for sustainable lithium sourcing, while consumers can contribute by supporting technologies and products that prioritize environmental stewardship.

Moreover, education and public awareness play pivotal roles in shaping the lithium narrative. By understanding the intricacies of lithium's journey, individuals can make informed decisions about their energy use and advocate for policies that promote sustainable practices. As the demand for lithium continues to surge, fostering a well-informed society is key to balancing technological progress with environmental responsibility.

The transformative power of lithium and its integral role in the battery revolution underscores a broader narrative: the need for a collective commitment to sustainability. From the initial ore extraction to the electrification of our vehicles and devices, each step requires careful planning, innovation, and ethical considerations. Embracing this journey isn't just about advancing technology; it's about securing a sustainable future for generations to come.

The case of lithium offers a glimpse into the broader landscape of critical resources that will shape our world. As we delve deeper into the realms of sustainable technologies, the lessons learned from lithium's journey guide us in building a future where technological advancements align with environmental preservation. By championing responsible resource management and fostering innovation, we can navigate the challenges and seize the opportunities that lie ahead, creating a world where technology and sustainability coexist harmoniously.

Environmental Implications and Alternatives are increasingly under the spotlight as our dependency on lithium grows. With the battery revolution sweeping through industries and homes, understanding the environmental consequences and exploring alternative solutions is crucial. To paint a detailed picture, we must scrutinize both the extraction processes and the potential paths towards more sustainable practices.

The extraction of lithium is not without significant environmental costs. Whether sourced from hard rock mining or lithium-rich brines, both methods entail considerable ecological footprints. For instance, hard rock mining results in substantial land disruption and landscape alteration. One can't ignore the aftermath of deforestation and the subsequent loss of biodiversity in mining areas.

Moreover, mining affects local water resources dramatically. In arid regions, where lithium brine extraction is prevalent, vast amounts of water evaporate during the process, impacting local communities and ecosystems that rely on those water sources. The depletion of these critical resources raises concerns about the long-term sustainability of lithium extraction.

The environmental strain isn't limited to the point of extraction. Processing lithium into battery-grade material involves chemicals that can lead to soil and water contamination if not managed properly. This contamination not only harms local wildlife but also poses a risk to human health. Therefore, the entire lifecycle of lithium must be examined comprehensively to develop a more sustainable pathway.

In seeking alternatives, one promising avenue is the recycling and reuse of lithium-ion batteries. The current recycling rates for these batteries are meager, but with the rise of electric vehicles, the importance of recycling gains even more prominence. Technological advancements in battery recycling promise to recover a significant proportion of lithium, reducing the need for virgin material extraction.

However, the recycling process isn't free from challenges. Efficiently retrieving lithium, cobalt, and other valuable metals from used batteries requires sophisticated technologies and robust systems to handle the influx of end-of-life products. Establishing a closed-loop system where batteries are continuously recycled can alleviate some environmental burdens.

Another alternative worth exploring is the development of new battery chemistries. Researchers are investigating sodium-ion, magnesium-ion, and solid-state batteries as potential substitutes for lithium-ion systems. These alternatives could offer comparable energy densities and better safety profiles, all while relying on more abundant resources. Advancements in these areas could dramatically reduce reliance on lithium and its associated environmental impacts.

Consider, too, the potential for improving mining practices. Innovations in green mining techniques aim to reduce the environmental footprint of resource extraction. Such methods include the use of less water-intensive processes, reforestation of mined areas, and the introduction of more stringent environmental regulations. These practices can significantly mitigate the adverse effects of mining activities.

One compelling solution lies in the utilization of Artificial Intelligence (AI) and Big Data technologies. AI can optimize mining operations to minimize resource use and environmental impact, while Big Data analytics provide valuable insights for improving efficiency and sustainability practices throughout the supply chain. These technological tools are pivotal in ensuring a more responsible approach to resource management.

To fully grasp the environmental implications, one must also consider the social aspect. Ethical considerations come into play when local communities face displacement or livelihoods are disrupted by mining activities. Sustainable extraction practices should prioritize the well-being of these communities, ensuring that they are part of the decision-making process and benefit from the economic activities resulting from mining.

In conclusion, addressing the environmental implications of lithium extraction and considering alternatives requires a multifaceted strategy. While recycling, new battery chemistries, and innovative mining techniques offer promising solutions, it is imperative to incorporate ethical and technological advancements to formulate comprehensive sustainability practices. The urgency for real change drives us to rethink our approach to resource management, with a critical need for global cooperation and innovation.

As we look towards building a sustainable future, it is essential to balance the need for advanced energy storage with the environmental and social responsibilities that come with it. By integrating recycling efforts, exploring alternative materials, and harnessing cutting-edge technologies, we can envision a world where both technological progress and environmental stewardship coexist harmoniously.

Chapter 3:
Harnessing the Power of the Sun

Transitioning to renewable energy isn't just a dream; it's an imperative. Solar power stands at the forefront of this transformation, offering an inexhaustible and clean energy source. Yet, the promise of harnessing the power of the sun requires substantial innovation in solar energy materials and technology. From the dawn of silicon-based solar cells, we've now moved into the age of perovskite solar cells, which promise higher efficiencies and lower production costs. However, generating solar energy is only part of the equation; efficient storage solutions are pivotal in bridging the gap between energy generation and consumption. Technological breakthroughs in battery storage and other energy retention systems are crucial for ensuring that the power captured from sunlight can be reliably utilized whenever needed. As we harness this abundant resource, we're not just generating power; we're charting a path towards a sustainable and resilient future, advancing towards a world where solar energy becomes a cornerstone of our global energy infrastructure.

Solar Energy Materials and Innovations

The advancement of solar energy technology hinges on the continuous evolution of materials and innovations that enhance efficiency and sustainability. Silicon-based solar cells have dominated the market for decades, but they are gradually being complemented and even supplanted by next-generation materials like perovskites, which promise higher efficiency rates and lower production costs. Additionally, research into organic photovoltaic cells opens the door for flexible, lightweight solar panels that can be integrated into various surfaces, from building facades to wearable devices. Such innovations are not just about increasing energy conversion rates; they're about rethinking how solar energy can be harvested and integrated into daily life. By focusing on materials that are abundant, non-toxic, and easy to recycle, the solar industry is contributing to a more sustainable future. These breakthroughs are crucial for creating a resilient energy infrastructure that can reliably support global needs, inspiring a new era of clean energy adoption. As we embrace these technological advancements, we move one step closer to a world where solar energy plays a pivotal role in powering our societies.

Next-Generation Solar Panels: Perovskites and Beyond are capturing the imagination of scientists and environmentalists alike. Imagine a world where solar panels are not just more efficient but also more versatile and affordable. That's the promise held by perovskite solar cells.

Perovskites have an extraordinary ability to absorb light, making them a game-changer in solar technology. Unlike traditional silicon-based panels, perovskites can be tuned to capture different wavelengths of light more efficiently. This adaptability translates into higher energy conversion rates, edging us closer to optimizing solar energy into our everyday lives.

But what exactly are perovskites? At their core, perovskites are a group of materials sharing a unique crystal structure. Their practical benefits extend beyond just solar energy; their applications range from LED lights to various sensing technologies. The appeal of perovskites lies in their compositional flexibility, which opens up a plethora of possibilities for innovation.

Traditional silicon panels, while revolutionary, come with significant drawbacks. The manufacturing process of silicon solar panels is energy-intensive and costly. In contrast, perovskite solar cells offer a simpler, energy-efficient, and less expensive alternative. Manufacturing perovskite cells involves a solution-based process that can be conducted at relatively low temperatures, reducing both environmental impact and production costs.

One can't help but get excited about the potential for perovskite solar cells to revolutionize where and how we deploy solar energy solutions. For example, their flexibility allows them to be integrated into a variety of surfaces, not just rigid solar panels. Imagine solar roofing tiles, window coatings, and even clothing that generates electricity — all could soon be possibilities.

However, the road to commercializing perovskite solar cells isn't without its challenges. Stability remains a significant hurdle. Prolonged exposure to moisture, oxygen, and ultraviolet light can degrade perovskite materials. Researchers are fervently seeking ways to improve the longevity of these cells, experimenting with protective coatings and novel material compositions to enhance durability.

When discussing next-generation solar technologies, it would be remiss not to consider the sustainability of raw materials. One of the compelling aspects of perovskites is their material efficiency. They require less raw material compared to silicon-based panels, potentially mitigating some environmental concerns associated with large-scale solar production.

Innovation in perovskites isn't happening in isolation. It's part of a larger movement towards hybrid solar cells that combine the best properties of different materials. For instance, tandem cells that layer perovskites on top of silicon or other substrates are showing remarkable efficiencies. These layered structures aim to leverage the advantages of each material to maximize the overall energy output.

The adoption of these next-generation solar panels could profoundly affect global energy dynamics. By making solar power even more affordable and accessible, we could democratize energy access, especially in developing regions where traditional infrastructure is lacking. Microgrids employing manageable, cost-effective perovskite panels could provide reliable electricity to remote communities, fostering social and economic advancements.

But the promise of perovskites isn't just about energy; it's about imagining a future where sustainable technologies lead the way to a cleaner planet. Each scientific breakthrough in this field brings us a step closer to a world less reliant on fossil fuels and more attuned to the renewable capabilities provided by natural resources.

Beyond perovskites, researchers are also eyeing other materials that could further propel solar technology. Emerging substances like organic photovoltaics (OPVs) and quantum dots promise even more versatility and efficiency. OPVs offer the potential for ultra-light, flexible solar panels, while quantum dots could lead to significant advancements in energy conversion efficiencies through novel photoelectric effects.

The continuous evolution of solar technologies punctuates a broader narrative: humans have an incredible capacity for innovation in the face of environmental challenges. By pushing the boundaries of what's possible with next-generation materials, we not only pave the way for technological advancements but also embody the essence of a resilient, forward-thinking society.

Looking ahead, it's imperative for policymakers, investors, and community leaders to recognize and support these advancements. Substantial investments in research and development, paired with incentives for sustainable practices, will be crucial in translating laboratory successes into real-world applications.

As we continue this journey toward a sustainable future, the story of perovskites and beyond serves as a testament to human ingenuity. Harnessing the power of the sun more effectively could indeed be one of the defining achievements of our time. So, let's embrace these innovations, fostering a world where sustainability and technology walk hand in hand, ensuring a brighter future for generations to come.

Bridging the Gap: Storage Solutions for Solar Energy

Solar energy holds immense potential as a sustainable and virtually limitless power source. However, harnessing the sun's power isn't just about capturing sunlight efficiently; it's also about effectively storing that energy for use when the sun isn't shining. This brings us to a critical juncture in solar energy development: the storage solutions that bridge the gap between energy production and consumption.

Effective storage solutions for solar energy are essential for overcoming one of the fundamental challenges of renewable energy systems: their intermittency. The sun doesn't shine at night, and cloudy days can significantly reduce the power generated by solar panels. To maintain a continuous and stable energy supply, we need robust storage systems that can store excess energy produced during sunny periods and release it when needed.

Batteries are currently the most common form of energy storage for solar power systems, both at the residential and industrial scales. Modern lithium-ion batteries, for instance, have revolutionized energy storage with their relatively high efficiency, long cycle life, and decreasing costs. These batteries have the advantage of being scalable— ranging from small units for individual homes to large installations for grid-scale storage.

Yet, lithium-ion batteries are not without their drawbacks. Issues such as resource scarcity, potential environmental impact of extraction, and limited recycling capabilities pose significant challenges. This has spurred research into alternative materials and battery technologies that could offer more sustainable and efficient solutions. Solid-state batteries, which use a solid electrolyte instead of a liquid one, are one such promising technology. They potentially offer higher energy densities and better safety profiles compared to traditional lithium-ion batteries.

Flow batteries are another emerging technology with great promise for solar energy storage. Unlike conventional batteries, flow batteries store energy in liquid electrolytes contained in external tanks. This means their energy capacity can be easily scaled up by increasing the size of the tanks, making them particularly suitable for large-scale energy storage applications. Vanadium redox flow batteries and zinc-bromine batteries are among the notable examples currently being developed and deployed.

In addition to chemical battery storage, mechanical energy storage solutions such as pumped hydro storage (PHS) and compressed air energy storage (CAES) are gaining attention. PHS, which involves storing excess energy by pumping water to a higher elevation and releasing it through turbines when needed, is already widely used and represents the largest form of energy storage worldwide. CAES works by compressing air into underground caverns during periods of excess energy production and releasing it to drive turbines and generate electricity when demand is high.

Thermal energy storage (TES) is another innovative solution, particularly relevant for solar thermal power plants. TES systems store excess thermal energy in materials such as molten salts during the day. This stored heat can then be used to produce electricity at night or during cloudy periods, providing a reliable energy supply. The efficiency and cost-effectiveness of these systems are continually being improved, making TES a viable option for large-scale solar power plants.

Integrating these diverse energy storage technologies into our energy infrastructure requires smart grid systems capable of managing and distributing energy efficiently. Advances in artificial intelligence and machine learning are proving instrumental in this regard. These technologies can optimize the operation of energy storage systems, predict energy production and consumption patterns, and facilitate the seamless integration of stored energy into the grid.

Moreover, policies and incentives play a crucial role in promoting the adoption of energy storage solutions. Governments and regulatory bodies worldwide are increasingly recognizing the importance of energy storage for grid stability and renewable energy integration. Subsidies, tax incentives, and regulatory reforms are being implemented to encourage investment in energy storage technologies.

Community energy storage (CES) systems represent one innovative approach to decentralize and democratize energy storage. CES systems involve placing moderate-sized storage units within communities, enabling local energy sharing and enhancing grid resilience. This model not only empowers communities to manage their own energy needs but also helps in balancing the local grid and reducing transmission losses.

Economic considerations are inevitably a major factor in the deployment of energy storage solutions. While costs for technologies like lithium-ion batteries have been falling, further economic incentives are needed to make advanced storage systems more widespread. The development of cost-effective and high-performance materials will be key to reducing overall storage costs.

To achieve a sustainable and energy-secure future, collaboration between various stakeholders is essential. This includes researchers, technology developers, policymakers, and consumers. Successful energy storage solutions will emerge from a synergy of technological innovation, robust policy frameworks, and widespread public adoption.

The role of energy storage extends beyond just balancing supply and demand; it is vital for enhancing energy security and creating resilient energy systems capable of withstanding natural and anthropogenic disruptions. The integration of storage into the energy mix provides a buffer against energy price volatility and supply disruptions, contributing to a more stable and predictable energy market.

It's important to remember that energy storage solutions are not a one-size-fits-all proposition. Different regions and applications will require tailored approaches that take into account local resource availability, energy needs, and economic conditions. The diversity of storage technologies available allows for such customized solutions, ensuring that solar energy can be harnessed and utilized effectively in various contexts worldwide.

In conclusion, bridging the gap between solar energy production and consumption through advanced storage solutions is a pivotal step toward a sustainable energy future. By addressing the challenges of intermittency and enhancing grid resilience, we can unlock the full potential of solar energy and pave the way toward a cleaner, greener planet.

Chapter 4:
Wind Energy: Materials
for the Invisible Force

In the grand vision of a sustainable energy future, wind energy stands tall, harnessing the relentless power of nature's own breath. Transforming this invisible force into tangible energy requires an intricate ballet of materials science and engineering innovation. Modern wind turbines are marvels of design and engineering, utilizing composites like fiberglass and carbon fiber to create blades that are both incredibly strong and lightweight. These materials allow turbines to capture the kinetic energy of the wind more efficiently, translating it into electrical power that's fed into our grids and homes. Whether offshore, where the winds are wild but the conditions harsh, or onshore, where accessibility and land usage present their own challenges, the materials chosen play a crucial role. As we advance, the blend of traditional alloys, cutting-edge composites, and innovative coatings will dictate the sustainability and efficiency of wind energy systems. In this way, the quiet hum of a spinning turbine becomes a powerful testament to human ingenuity, potentially reshaping our energy landscape with a force that's as invisible as it is unstoppable.

Advancements in Turbine Technology

The transformation in turbine technology is pivotal to maximizing wind energy's potential in our quest for sustainability. We've seen groundbreaking improvements in rotor blade design, allowing turbines to capture wind more efficiently by using advanced composites that enhance durability and reduce weight. This, in turn, minimizes material fatigue and increases the lifespan of each turbine. Additionally, breakthroughs in aerodynamic modeling have led to the development of larger turbines capable of generating more power from slower wind speeds, a crucial factor in diverse geographic locations. Coupled with innovations in digital twin technology and predictive maintenance, the operational efficiency of wind farms is reaching unprecedented levels. These advancements not only contribute to reducing the levelized cost of energy but also align seamlessly with global efforts to transition to cleaner energy sources. It's evident that the future of turbine technology holds promise for significant strides in energy sustainability and offers a beacon of hope for achieving our global renewable energy targets.

The Role of Composites in Wind Energy cannot be overstated as we transition to a more sustainable energy system. Wind energy has gained considerable traction over the last two decades, definitely evolving as a cornerstone in renewable energy strategies worldwide. One of the key enablers of this growth has been the use of advanced composite materials in wind turbine manufacturing.

Composites are a game-changer in creating high-performance wind turbines. These materials, typically composed of fibers embedded in a resin matrix, offer an excellent combination of strength, flexibility, and lightweight properties. Such properties are essential for the construction of wind turbine blades, which must be both robust and aerodynamic.

Consider the sheer size of modern wind turbine blades, often spanning over 80 meters for offshore applications. The use of composites allows for these massive structures to be lightweight enough to rise to such heights while maintaining the necessary rigidity to withstand varying wind conditions. Traditional materials like metals simply can't offer the same balance of attributes without significantly increasing weight and cost.

The most common composites used in wind energy are glass fiber-reinforced polymers (GFRP) and, increasingly, carbon fiber-reinforced polymers (CFRP). GFRP offers a good balance of cost and performance, making it the go-to material for many manufacturers. CFRP, while more expensive, is favored for its superior strength-to-weight ratio and fatigue resistance, qualities that are particularly beneficial for the longer blades used in offshore wind farms.

Harnessing wind energy on a large scale presents unique challenges — notably, the turbines must operate efficiently over long periods, often in harsh environments. Composites excel in this arena thanks to their durability and resistance to environmental factors such as corrosion and UV radiation. This longevity lessens the frequency and cost of maintenance, making wind farms more economically viable.

The environmental benefits of these materials shouldn't be overlooked. Composites offer a pathway to creating greener wind turbine components by reducing the overall carbon footprint during manufacturing and operational phases. Unlike metal counterparts that require energy-intensive mining and production processes, composites can be fabricated with comparatively lower energy inputs. This minimal energy expenditure aligns well with the overarching goals of sustainability.

Furthermore, the adaptability of composites makes them ideal for innovative design methodologies. With the use of advanced computer modeling and simulations, engineers are able to tailor the composite layups to optimize blade performance under various wind conditions. This flexibility is crucial for achieving the efficiency gains necessary to make wind energy more competitive with fossil fuels and other renewable sources.

There is, of course, room for improvement. Issues such as the recyclability of composite materials are currently under the spotlight. The very properties that make composites so effective for wind turbine blades also complicate their end-of-life disposal. Research and development efforts are increasingly focused on creating more sustainable composite alternatives and improving recycling processes.

For example, efforts are being made to develop thermoplastic composites that can be more easily recycled compared to traditional thermosetting composites. This innovation could substantially mitigate the waste issues associated with wind turbine decommissioning, pushing the envelope further towards a circular economy for renewable energy technologies.

A tightrope is often walked between performance and sustainability, but composites in wind energy demonstrate that these goals can converge. As the demand for renewable energy surges, the role of composites will only become more pronounced. This sector promises further breakthroughs that will enhance the performance, reduce the costs, and mitigate the environmental impacts associated with wind turbines.

The wind energy sector is a compelling case study of how material science can drive sustainable innovation. As we look toward the future, it's evident that composites will continue to play a pivotal role. They embody the spirit of ingenuity and resourcefulness needed to meet the world's energy demands without compromising ecological balance.

As public and private stakeholders mobilize around climate goals, the integration of composite materials in wind energy solutions is a testament to human ingenuity. These advanced materials provide tangible benefits that enable wind energy to be a reliable and sustainable pillar of our energy system. By investing in composites research and application, we are investing in a cleaner, more resilient future.

In this transformative period, composites not only help us harness the invisible force of the wind more effectively but do so in a way that aligns with the principles of sustainability. They exemplify how thoughtful material selection and technological innovation can drive the transition towards a more sustainable world.

Ultimately, the role of composites in wind energy echoes a broader theme: the necessity of marrying technological advancement with environmental stewardship. As we stand at the precipice of a renewable revolution, the continued evolution and refinement of composite materials will be instrumental in ensuring that wind energy remains a key player in the global shift towards a more sustainable future.

Offshore vs. Onshore: Material Challenges and Opportunities

As wind energy becomes an integral part of our path towards a sustainable future, the distinction between offshore and onshore wind farms emerges as a focal point of discussion. Each offers its own set of opportunities and challenges, especially from a materials perspective. Understanding these differences can guide innovation and investment in ways that maximize efficiency and minimize environmental impact.

In the expansive canvas of offshore wind farms, space is abundant, allowing for the installation of larger turbines that can harness stronger and more consistent winds prevalent over the ocean. However, this comes at a cost, not just financial but material. The harsh marine environment demands materials that can withstand saline corrosion, high moisture levels, and the relentless forces of waves and wind.

Corrosion-resistant alloys and **protective coatings** are pivotal for the longevity and reliability of offshore structures. Stainless steel and various composite materials have shown promise, but they are not without their own sets of challenges. High costs and complex manufacturing processes often accompany these advanced materials, posing a significant barrier to widespread adoption.

In contrast, onshore wind farms, while more accessible and easier to maintain, face their own material challenges. Turbines located on land must contend with variable wind speeds and differing geographic conditions. This necessitates a nuanced approach to material selection and turbine design, ensuring they are adaptable and robust enough to perform efficiently in a diverse range of environments.

The foundation materials for onshore turbines are another crucial consideration. While concrete and steel are the go-to choices, ongoing research is exploring more sustainable alternatives. Innovations like carbon-reinforced concrete and new alloy compositions aim to reduce the carbon footprint without compromising structural integrity.

Interestingly, the transportation and installation of wind turbines also highlight stark differences. Offshore installations often require specialized ships and complex logistics, from transporting turbine components to installing them securely on the seabed. This logistics puzzle accentuates the need for lightweight but sturdy materials, which can reduce transportation costs and simplify assembly processes.

Onshore turbines, on the other hand, benefit from relatively easier transportation logistics but face restrictions related to road sizes and weight limits. This drives the demand for modular components and materials that strike a balance between weight efficiency and durability.

One can't overlook the role of composites in both offshore and onshore wind technologies. Composites offer excellent strength-to-weight ratios, making them ideal for the massive blades and intricate nacelle structures. Innovations in composite materials - such as the development of thermoplastic composites - promise easier recycling and lower environmental impacts, which are crucial for the sustainable expansion of wind energy.

The quest for efficiency and sustainability doesn't end with the installation of turbines. Maintenance and repair are critical, particularly for offshore turbines where access can be difficult and costly. Materials that minimize wear and tear, and innovative solutions like self-healing polymers, could revolutionize maintenance practices, reducing downtime and extending the operational life of turbines.

Moreover, the decommissioning phase brings its own set of material challenges and opportunities. Responsibility for the lifecycle of wind turbine materials, from cradle to grave, is becoming a focal point. Research into recyclable materials and designs that facilitate easy disassembly and material recovery is paramount for minimizing the environmental footprint of wind energy projects.

Despite these challenges, the opportunities in both offshore and onshore wind infrastructure are vast. The development of advanced materials not only promises to overcome current limitations but can also unlock new potentials. For example, biomaterials and biodegradable composites are on the horizon, offering the possibility of turbines that leave no lasting footprint once their life cycle is complete.

Collaboration across industries is essential. Innovations in materials science need to be coupled with advancements in manufacturing technologies, logistics, and maintenance strategies. By fostering a multidisciplinary approach, we can ensure that wind energy not only meets our current needs but is also prepared for the demands of future generations.

Fundamentally, the drive towards more efficient and sustainable wind energy production is a testament to our ability to blend environmental consciousness with technological innovation. Whether harnessing the relentless winds offshore or optimizing onshore installations, the journey is replete with both challenges and boundless opportunities for a better, greener future.

The success of wind energy technology hinges on our ability to continue pushing the boundaries of materials science. By addressing the intricate and often contrasting needs of offshore and onshore wind farms, we can craft solutions that are not only technically sound but also environmentally and economically sustainable.

As we venture forward, the lessons learned from our experiences in both terrains will guide us towards smarter, more resilient designs. The marriage of material innovation with pragmatic engineering will create a pathway not just for wind energy but for a wider spectrum of sustainable technologies that define the future of energy landscapes.

The horizon for wind energy is bright. Embracing the material challenges and seizing the opportunities presented by both offshore and onshore wind farms will propel us towards a future where renewable energy isn't just viable, but indispensable.

Chapter 5:
The Water-Energy Nexus

The intricate relationship between water and energy, known as the water-energy nexus, is a fundamental yet often overlooked component of our sustainable future. Water is indispensable for energy production, whether it involves cooling power plants, extracting oil, or generating hydroelectric power. Conversely, producing potable water and treating wastewater require significant amounts of energy, creating a symbiotic dependency that underscores the urgency of addressing both resources in tandem. Innovations, such as advancements in turbine materials for hydro and tidal power and cutting-edge desalination technologies, offer promising pathways to tackle the dual challenge of water scarcity and energy demands. Nonetheless, the path to harmonizing these critical resources is a complex endeavor fraught with both technological and policy-driven considerations. As we probe into the depths of these challenges, it becomes increasingly clear that the water-energy nexus is not just an environmental issue—it's a pivotal battleground for our sustainable development, urging a paradigm shift towards integrated resource management.

Hydro and Tidal Power: The Potential of Water

Water's potential to power our world is vast and largely untapped, shining as one of the most promising energy sources for a sustainable future. Harnessing the power of flowing rivers and ocean tides offers not just a reduction in our reliance on fossil fuels but can also complement other renewable energy systems by providing consistent and reliable power. The ingenuity behind hydroelectric dams has been well-documented, yet it's the innovative tidal and wave energy technologies that are beginning to turn heads. These newer technologies promise to transform our coastal regions into hubs of renewable energy production, capitalizing on the relentless and predictable movements of the sea. Investing in advanced turbine materials and designs that withstand corrosive aquatic environments will be crucial. More than just a technical endeavor, the move to hydro and tidal power embodies a broader shift towards embracing the natural dynamics of Earth's water systems as benevolent forces for energy generation. As we face the imperative of climate change, tapping into the kinetic splendor of water should not just inspire us—it should galvanize us into action.

Innovations in Turbine Materials and Design represent a key frontier in both energy generation and sustainability. Turbines—whether harnessing the power of wind, water, or even steam—are crucial elements in our quest for clean energy. However, the traditional materials and designs have often faced limitations in efficiency, durability, and environmental impact. That's where innovative materials and clever design principles come into play, challenging the status quo and leading us towards a more sustainable future.

Turbine blades are the heart of energy conversion in wind and hydro power systems. Conventional materials like steel and aluminum have long been the backbone of these machines. However, they come with downsides: weight, susceptibility to corrosion, and substantial maintenance needs. Enter composite materials. Composites, primarily made of carbon fiber and glass-reinforced polymers, are game-changers. Not only do they offer significant weight reduction, but they also boast heightened durability and resistance to environmental stressors like saltwater in offshore wind farms. This is vital for longevity and efficiency.

Imagine a wind turbine blade so light and resilient that it can withstand the harshest environmental conditions with minimal maintenance. That's what composite materials promise. But we're not just looking at traditional composites; innovative research is pushing boundaries with hybrid composites, integrating different types of fibers to tailor properties for specific applications. This approach maximizes performance while optimizing cost, giving us a peek into the future of smart materials.

Another groundbreaking development in turbine materials is the advent of shape-memory alloys (SMAs). These metals have the unique ability to return to a predefined shape when heated. In the context of turbines, SMAs could be employed to create adaptive blades that change their shape in response to varying wind or water conditions. This adaptability can significantly increase the efficiency of energy capture, making turbines more versatile and efficient than ever before.

Moreover, innovations aren't limited to the blade materials alone. The design and materials used in the structural components of turbines have seen transformative changes as well. Engineers are exploring modular turbine designs that use pre-fabricated, snap-together components. This could revolutionize the assembly process, making it faster and more cost-effective, and reducing the carbon footprint associated with construction and maintenance. Imagine assembling a massive wind turbine in a fraction of the time it currently takes, with far less environmental impact.

Composite and shape-memory materials continue to make strides, but another material waiting in the wings to reshape turbine technology is graphene. Known for its unmatched strength and electrical conductivity, graphene is being explored for use in various turbine components. Graphene-coated blades could dramatically reduce wear and tear, potentially leading to maintenance-free turbines. Although still in the research phase, the applications of graphene in turbine technology could be revolutionary.

But it's not just all about new materials; innovation in design plays an equally critical role. Bionic design principles, inspired by nature, are pushing the envelope in turbine technology. By mimicking the structures found in bird wings and fish fins, engineers have created blades that are quieter, more efficient, and better suited for varying environmental conditions. These biomimetic designs offer a blend of elegance and efficiency, allowing turbines to operate more harmoniously within their ecosystems.

Additionally, engineers are playing with geometry to enhance turbine efficiency. Variable geometry turbines, where blade angles can be adjusted in real-time, have shown significant efficiency improvements. This dynamic approach ensures that turbines can operate optimally across a wider range of wind speeds and water flows, extracting more energy while reducing wear and tear. It's a step towards creating more intelligent and responsive energy systems.

Moving on to construction methods, 3D printing is making waves in turbine technology. Additive manufacturing allows for the creation of complex, optimized components that traditional methods can't achieve. This not only reduces material waste but also opens the door for rapid prototyping and customization. Imagine on-site 3D printing of turbine parts, minimizing logistical challenges and accelerating deployment timelines. The future of rapid, eco-friendly turbine assembly might be closer than we think.

Enhancements in software and computational modeling also play a crucial role in design innovation. Advanced simulations enable engineers to model every aspect of turbine performance before manufacturing even begins. This approach minimizes costly trial and error, ensuring that every turbine is as efficient and reliable as possible from the outset. It's like giving engineers the ability to predict the future, refining their designs in silico before they ever hit the real world.

In the realm of hydro power, materials and design innovations are equally compelling. New developments in non-corrosive materials can vastly improve the lifespan of underwater turbines. This investment in longevity reduces long-term costs and environmental disturbances associated with frequent maintenance. Moreover, innovative turbine shapes that mimic marine life can reduce harm to aquatic organisms, marrying technology and environmental stewardship.

The conversation about innovative turbine materials and design wouldn't be complete without mentioning recycling. Advanced recycling techniques ensure that at the end of their life cycle, turbine components don't end up as landfill waste. Eco-friendly material choices and modular designs facilitate easy disassembly and recycling, feeding back into the circular economy we so desperately need. It's a cradle-to-cradle approach, making the lifecycle of turbines as green as their purpose.

While we envision these futuristic materials and designs, the role of policy and international cooperation can't be ignored. Supportive policies that endorse research and development, as well as international collaboration, are pivotal for accelerating these innovations from the lab to real-world applications. Together, these efforts create an environment where pioneering ideas can thrive, ensuring that turbine technologies evolve in tandem with our sustainability goals.

In summary, the future of turbine technology is brimming with potential. From composites and SMAs to graphene and bionic designs, the innovations in turbine materials and design are making renewable energy not just viable but exceptionally efficient and sustainable. These advances are setting the stage for a future where our energy systems are smarter, more resilient, and kinder to the planet. It's an invigorating glimpse into what's possible when ingenuity meets purpose, forging a path towards a sustainable and empowered world.

Desalination Technologies: Quenching the World's Thirst

Water is one of the most fundamental resources that sustain life. Yet, a significant portion of the world's population struggles with water scarcity. As our global community continues to grow, so does the demand for fresh water, putting immense pressure on available resources. This brings us to the intricate dance between water and energy, a nexus that underpins much of our current technological advancement and future progress.

Desalination, the process of removing salt and other impurities from seawater, has emerged as a powerful contender in addressing the burgeoning water crisis. By transforming saline seawater into potable freshwater, desalination holds the potential to alleviate water scarcity in arid regions and overpopulated urban centers. The technology, however, doesn't come without its challenges, primarily due to its high energy consumption. As we explore the various desalination technologies, it's essential to consider their sustainability and impact on global energy resources.

There are several different methods of desalination, each with its own set of advantages and limitations. The most commonly employed techniques are thermal distillation and membrane-based processes like reverse osmosis (RO). Thermal distillation involves heating seawater to create steam, which is then condensed back into liquid water, leaving the salt behind. While effective, this process is energy-intensive, often requiring significant amounts of fuel to produce the necessary heat.

On the other hand, reverse osmosis is currently the most popular method due to its relatively lower energy requirements and efficiency. It works by forcing seawater through a semi-permeable membrane that blocks salt and other larger molecules. While RO has revolutionized the desalination space, it still faces challenges, particularly in handling the concentrated brine byproduct and the fouling of membranes.

Emerging technologies seek to address these challenges and make desalination more energy-efficient and environmentally friendly. One such innovation is forward osmosis (FO), which utilizes a natural osmotic pressure gradient to draw water through a membrane, theoretically requiring less energy than reverse osmosis. Though promising, FO is still in its nascent stages and requires further development to be a viable large-scale solution.

Another cutting-edge approach is the integration of renewable energy sources with desalination plants. Using solar or wind energy to power these facilities can significantly reduce their carbon footprint and operating costs. Solar desalination, for instance, directly harnesses solar heat to evaporate water, eliminating the need for external fuel sources. Similarly, wind energy can drive the pumps and machinery necessary for membrane-based processes, making the entire system more sustainable.

In addition to technological improvements, addressing the issue of water and energy from a holistic perspective is crucial. Advances in materials science, such as the development of more durable and efficient membranes, play a pivotal role. Researchers are continually exploring novel materials like graphene, known for its exceptional permeability and strength, to enhance the efficiency of desalination membranes.

Despite these advancements, the economics of desalination remain a significant hurdle. The high capital and operational costs associated with setting up and maintaining desalination plants can be prohibitive, especially for developing nations. This underscores the importance of international cooperation and funding in accelerating the adoption of desalination technologies. Financial models that promote investment in sustainable water resources and incentivize the use of renewable energy can drive progress in this critical field.

Moreover, the environmental impact of desalination can't be overlooked. The process of extracting and returning brine into the ocean can disrupt local marine ecosystems. Effective management of brine discharge and the exploration of beneficial uses for concentrate, such as in mineral recovery, are essential steps toward minimizing ecological harm.

Public perception and acceptance play a vital role in the implementation of desalination projects. Community engagement and transparent communication about the benefits and potential drawbacks can foster support for these initiatives. Educating the public on the importance of sustainable water management and the role of desalination can drive collective action and advocacy for responsible water use.

The integration of desalination into the broader framework of water management involves not just technical, but also policy-level considerations. Policymakers must create robust frameworks that support research, deployment, and maintenance of desalination systems. This includes setting strict environmental regulations to mitigate impacts and providing subsidies or incentives for renewable energy integration.

Looking ahead, the future of desalination could be shaped by ground-breaking advancements such as biomimetic membranes that mimic the natural desalination processes found in living organisms. These membranes offer the potential for highly efficient filtration with significantly reduced energy requirements. The key lies in translating these innovative concepts from the laboratory bench to industrial-scale applications.

Desalination is far from a silver bullet that will single-handedly resolve the global water crisis. It must be part of a multifaceted strategy that includes water conservation, efficient usage, and recycling. However, as we strive toward sustainable management of our water resources, desalination represents a crucial technology in expanding the availability of fresh water where it is most needed.

In conclusion, desalination technologies hold significant promise for quenching the world's growing thirst. By continually advancing and innovating in this field, we can make desalination more efficient, cost-effective, and environmentally friendly. In doing so, we take a vital step towards securing water for future generations while balancing the delicate water-energy nexus.

In the next section, we will delve further into the role of bio-based and biodegradable materials in building a sustainable future. Understanding how nature-inspired solutions and renewable resources can revolutionize industries is key to creating a more sustainable and resilient world.

Chapter 6:
Bio-Based and Biodegradable
Materials

At the intersection of innovative technology and ecological stewardship lies the burgeoning field of bio-based and biodegradable materials. These sustainable alternatives to fossil fuel-derived products promise not only to reduce our environmental footprint but also to usher in a new era of material science. Bio-based materials, derived from renewable biological sources like plants and algae, are revolutionizing industries from packaging to automotive components. Meanwhile, biodegradable materials offer solutions that can significantly mitigate the burgeoning waste crisis. These materials break down more easily in natural environments, reducing long-term pollution and supporting circular economies. The rise of bioplastics signals a transformative shift, emphasizing that sustainable options don't have to compromise performance. In fact, with continuous advancements, they have the potential to outperform traditional materials while addressing the dire need for eco-friendly innovation. The momentum gained by these materials suggests an age where technology and nature collaborate seamlessly, propelling us toward a more sustainable future.

The Rise of Bioplastics

Bioplastics are emerging as a game-changer in the quest for sustainable materials. Unlike conventional plastics derived from fossil fuels, bioplastics are produced from renewable resources such as corn starch, sugarcane, and algae. These innovative materials not only reduce our dependency on petroleum but also offer the promise of biodegradability, minimizing environmental impact. With advancements in biotechnology, the efficiency of producing these materials is improving, making them more commercially viable. The significant reduction in greenhouse gas emissions during production compared to traditional plastics further bolsters their environmental credentials. As companies and consumers become more eco-conscious, the demand for bioplastics is expected to surge. This shift represents a critical step towards reducing the ecological footprint of our material consumption and fostering a more sustainable future. By embracing bioplastics, we're not just mitigating plastic pollution; we're paving the way for a circular economy where materials are reused and recycled, benefitting both the planet and society.

From Plant to Plastic: The Process Explained delves into the fascinating journey of turning renewable plant materials into versatile, eco-friendly plastics. As the shift away from fossil-fuel-based products accelerates, understanding this transformative process becomes paramount. Let's break it down to see how the magic happens.

It all starts with selecting the right plant materials. The most common sources are corn, sugarcane, and cellulose. These plants are chosen because of their high starch content and ability to be cultivated sustainably. Corn, for instance, can yield large amounts of glucose when broken down, which is vital for the subsequent steps in the bioplastic-making process.

The first major step is extraction. The plant biomass, whether it's from corn or sugarcane, is harvested and then milled to release the sugars. This raw material undergoes a process called saccharification, where enzymes break down the complex carbohydrates into simple sugars like glucose. This simplicity is what allows these sugars to be further manipulated and eventually polymerized.

Fermentation is the next critical phase. Here, microorganisms, typically bacteria or yeast, are introduced to the glucose-rich liquid. These microorganisms feed on the sugars, producing byproducts like lactic acid or ethanol. This biological conversion is essential because it transforms the simple sugars into monomers – the building blocks of polymers.

Once we have our monomers, such as lactic acid, the polymerization process begins. This involves chemically linking the monomers to form long, repeating chains, resulting in polylactic acid (PLA). This bioplastic is known for its biodegradable properties and is widely used in packaging and disposable items.

However, PLA is just one type of bioplastic. Others include polyhydroxyalkanoates (PHAs) and bio-polyethylene (bio-PE). Each has its unique manufacturing processes but shares the common foundation of plant-derived sugars converted through fermentation.

Purification and compounding follow. After polymerization, the bioplastic is purified to remove any residual monomers or byproducts. This purification ensures that the final product is consistent and safe for its intended use. The polymer pellets can then be compounded with additives to enhance properties like flexibility, strength, and durability.

From here, these purified, and often compounded, bioplastics are ready to be shaped. Traditional plastic manufacturing techniques, such as injection molding, extrusion, and blow molding, can be employed. This versatility in processing means that bioplastics can seamlessly replace conventional plastics in many applications without the need for new machinery or technology.

As the demand for sustainable materials grows, the production of bioplastics continues to evolve. Researchers are exploring the use of non-food plant sources, like algae and agricultural waste, to avoid competition with food resources. These advancements are not only crucial for sustainable production but also for tapping into the full potential of biodegradable materials.

That's not to say there aren't challenges. One of the primary hurdles is scalability. While the technology to produce bioplastics exists, scaling up to meet global demand while maintaining sustainable practices is no small feat. It requires concerted efforts in agriculture, biotechnology, and industrial processes to synchronize effectively.

Another significant consideration is the lifecycle analysis. It's imperative to evaluate the environmental impact of bioplastics from cradle to grave, encompassing everything from the cultivation of the plants to the disposal or recycling of the end product. Only through a comprehensive analysis can we truly gauge the sustainability of bioplastics compared to their petroleum-based counterparts.

Despite these obstacles, the potential benefits of plant-based plastics are immense. They offer a renewable alternative to depleting fossil fuel resources and propose a solution to the mounting plastic waste crisis. Moreover, by integrating bioplastics into the circular economy, we can move towards a more resilient and sustainable future.

The role of policy and consumer awareness in this transition cannot be overstated. Governments need to provide incentives for bioplastic production and adoption, while consumers must be educated on the benefits and proper disposal methods to maximize environmental gains.

In conclusion, the journey from plant to plastic is a testament to human ingenuity and the possibilities of sustainable innovation. By leveraging the remarkable capabilities of nature, we can forge new paths in material science that align with the planet's ecological boundaries. This process, intricate yet profoundly impactful, beckons us to reimagine how we produce and consume materials in a world increasingly defined by environmental stewardship.

Biofuels: Supporting a Sustainable Transportation Future

The momentum around biofuels is reaching new heights as the world collectively seeks alternatives to fossil fuels. These renewable energy sources promise not only to reduce carbon emissions but also to tackle the intricate puzzle of energy security and sustainability. As concerns about climate change grow louder, biofuels are emerging as a heroic, if somewhat complex, solution to an intricate problem.

Unlike traditional fuels derived from finite fossil resources, biofuels are produced from living organisms or organic waste. Common sources include plant materials, agricultural residues, and even algae. There are different types of biofuels, such as ethanol and biodiesel, each with its own set of production processes and applications. Their versatility makes them particularly appealing for a range of transportation methods, from cars and trucks to airplanes and ships.

An interesting aspect of biofuels is their potential to repurpose agricultural waste. For instance, corn stover, the leaves and stalks left in the field after harvest, can be transformed into ethanol. Similarly, used cooking oils and animal fats can be converted into biodiesel. This not only provides a substrate for fuel but also helps manage agricultural and organic waste more effectively.

A primary advantage of biofuels is their compatibility with existing internal combustion engines. This "drop-in" capability means that, in many cases, biofuels can be used without requiring significant modifications to vehicles or logistics infrastructure. This advantage cannot be overstated as it allows for a smoother, more manageable transition to greener energy sources.

However, the production of biofuels isn't without challenges. The debate surrounding the food versus fuel dilemma highlights a significant concern: should arable land be used to grow crops for fuel instead of food? Innovations in second-generation biofuels, which utilize non-food biomass, are increasingly addressing this issue. These innovations leverage advanced biochemical methods to convert lignocellulosic feedstocks, like wood and agricultural residues, into viable fuels.

Environmental sustainability is at the heart of the biofuels conversation. Ideally, biofuels should have a neutral or positive carbon balance. This means that the carbon dioxide absorbed by the plants during growth should offset the emissions produced during fuel production and combustion. Life-cycle assessments are crucial in evaluating the true environmental impact of biofuels, encompassing everything from cultivation and harvesting to transportation and fuel processing.

The development and scaling of biofuel technologies require significant investment, both financially and technologically. While first-generation biofuels made from food crops are relatively easy to produce, they offer limited sustainability. The next leap lies in third-generation biofuels, like those derived from algae, which promise higher yields and lower environmental footprints. These biofuels carry the potential to revolutionize the energy landscape, but they require further research and investment to become economically viable.

Take algae, for example. Algae can be grown in diverse environments, including non-arable land, and have a rapid growth rate. They are potent biofuel sources due to their high oil content. Moreover, algae can capture carbon dioxide from the atmosphere, potentially transforming them into both a biofuel source and a carbon sequestration tool. The promise is dazzling, but challenges remain in harvesting and processing these microscopic powerhouses at scale.

Investing in biofuel infrastructure, from advanced biorefineries to efficient distribution networks, is essential. Nations around the world are gradually recognizing the potential of biofuels to achieve their climate commitments. Mandates, subsidies, and tax incentives for biofuel production and usage are becoming more common, motivating industries to innovate and scale these technologies.

The societal benefits of biofuels extend beyond environmental sustainability. Developing countries, particularly those rich in biomass resources, stand to gain economically. Biofuel production can lead to job creation and promote rural development, addressing economic disparities while contributing to the global fight against climate change. Localized biofuel industries can enhance energy independence and reduce reliance on imported fossil fuels.

In the broader context of sustainable transportation, biofuels bridge a crucial gap. While electric vehicles (EVs) and hydrogen fuel cells represent the future of clean mobility, they require substantial changes in infrastructure and consumer behavior. Biofuels offer an immediate, interim solution to reduce emissions from the existing fleet of internal combustion engine vehicles, buying us valuable time as other sustainable technologies mature.

Consumer awareness and acceptance are pivotal. People need to understand the benefits and limitations of biofuels to support their adoption. Educational campaigns and transparent communication about the environmental, economic, and social impacts of biofuels can empower individuals to make informed choices, fueling demand for greener alternatives.

Looking to the horizon, it is clear that biofuels alone won't solve our transportation challenges. They are, however, a vital piece of the puzzle. As we transition to a more sustainable energy ecosystem, biofuels provide a bridge that allows us to make significant emissions reductions now while continuing to innovate for the future.

In conclusion, biofuels represent a dynamic and multifaceted approach to sustainable transportation. They offer the potential for significant environmental benefits, economic development, and improved energy security. By supporting research, fostering innovation, and building resilient biofuel industries, we take meaningful strides towards a sustainable transportation future. This journey demands collaboration across sectors, sustained investment, and, perhaps most importantly, a collective commitment to change.

Believing in the promise of biofuels and making concerted efforts to integrate them into our energy landscape can have lasting impacts. We can visualize a world where our reliance on fossil fuels diminishes, replaced by cleaner, renewable alternatives that support both our planet and our economy. The role of biofuels in this vision is not merely supplemental but fundamental as we drive towards a sustainable and resilient future.

Chapter 7:
The Future of Nuclear Power

As we transition to a more sustainable energy landscape, the future of nuclear power is perched on a transformative threshold. Harnessing the immense potential of nuclear energy requires innovation not just in technology, but in safety and environmental management too. Emerging technologies like thorium reactors show promise for cleaner, more efficient power generation, but the hurdles of nuclear waste mitigation remain formidable. Yet, advancements in material sciences and reactor design are paving the way for safer and more efficient systems. By addressing these challenges with both urgency and ingenuity, we can reimagine nuclear power as a pivotal component in our quest for a sustainable energy future, offering a balance of immense energy output with reduced environmental footprint.

Safe Nuclear Technology: Materials at the Forefront

Innovation in nuclear materials is setting the stage for safer and more efficient nuclear power. As we look to reduce carbon emissions and transition to cleaner energy, the development of advanced materials is crucial. Materials like high-performance alloys, advanced ceramics, and composite materials are being engineered to withstand extreme conditions within nuclear reactors, reducing risk and enhancing durability. Additionally, breakthroughs in cladding materials are significantly mitigating the risks of reactor failures and radioactive leaks. These advancements don't just promise safer energy—they pave the way for sustainable, long-term power solutions critical to our future. Investing in such technologies will bring us closer to a green, resilient energy grid, where nuclear energy plays a vital role alongside renewables. It's time to envision a future powered by not just the sun and wind, but also by safer, more efficient nuclear technology.

Thorium Reactors: A Cleaner Alternative are gaining recognition in the scientific community as a promising solution to the dual challenges of energy sustainability and environmental preservation. As we seek to address the escalating demand for energy without exacerbating climate change, thorium emerges as a compelling alternative to traditional nuclear fuels like uranium.

Thorium, a slightly radioactive metal, is abundant in the earth's crust, making it a more accessible resource compared to uranium. One kilogram of thorium can generate as much energy as 200 tons of coal, highlighting its potential in terms of energy density. With increasing global energy consumption, tapping into such a reliable and efficient resource could significantly reduce our dependence on fossil fuels.

The safety profile of thorium reactors is arguably one of their most attractive features. Unlike uranium reactors, thorium reactors operate in a manner that inherently prevents meltdowns. In a thorium reactor, if there is any malfunction or power cut, the reactor naturally cools down and solidifies, stopping the reaction without the risk of runaway reactions seen in uranium-based reactors.

Importantly, thorium reactors produce significantly less long-lived radioactive waste compared to their uranium counterparts. This reduction is crucial in addressing one of the most persistent problems associated with nuclear energy—the long-term storage and management of radioactive waste. By minimizing hazardous waste, thorium reactors present a more sustainable nuclear option that aligns with global environmental goals.

The potential of thorium reactors extends beyond safety and waste reduction. These reactors also have the capability to consume existing plutonium stockpiles. By incorporating thorium into the nuclear fuel cycle, we could not only mitigate the proliferation risks associated with plutonium but also transform the byproducts of previous nuclear activities into useful fuel.

Furthermore, thorium reactors can be designed as more efficient breeder reactors. Traditional reactors often use only a small fraction of the fuel, but thorium reactors are capable of generating more fissile material than they consume. This breeding process means that thorium fuels can be used more comprehensively, maximizing energy output while minimizing waste.

The technical feasibility of thorium reactors has been demonstrated over the decades, with various experimental reactors showing promising results. Notably, while the initial development costs of thorium reactors can be high, the long-term benefits in terms of reduced waste management and enhanced safety often outweigh these initial investments.

Public perception and policy play crucial roles in the advancement of thorium reactor technology. Greater public awareness and acceptance can lead to more significant investments and policy support, enabling the construction and operation of commercial thorium reactors. It is essential for policymakers to recognize the benefits and support the research and development needed to bring this technology to the forefront.

International collaboration can accelerate the deployment of thorium reactors. By sharing knowledge, resources, and best practices, countries can collectively overcome technical and economic barriers.

Collaborative efforts can help establish standardized safety protocols and regulatory frameworks, fostering a global environment conducive to the adoption of thorium-based nuclear power.

Educational initiatives are vital in promoting understanding and acceptance of thorium reactors. Universities and research institutions should incorporate thorium and its potential in their curricula, training a new generation of scientists and engineers who can lead the innovation and implementation of this technology.

Challenges remain, as with any emerging technology. Technical hurdles such as designing reactors that can efficiently utilize thorium and ensuring the economic viability of these reactors need continued research and development. However, the potential rewards for overcoming these challenges are immense—a safer, cleaner, and more sustainable energy future.

In light of ongoing climate change concerns and the need for a diversified energy portfolio, thorium reactors provide an attractive alternative that merits serious consideration. Their ability to offer a safer nuclear option with reduced environmental impact aligns perfectly with global sustainability goals.

To inspire action, we must illustrate the tangible benefits of thorium reactors to stakeholders, investors, and the general public. Effective communication strategies and public engagement can help build support for the necessary investments in thorium reactor research and infrastructure.

Ultimately, the transition to thorium reactors represents a step toward a more sustainable and resilient energy system. By embracing this technology, we can move closer to an era where our energy needs are met without compromising the health of our planet.

The shift to thorium-based nuclear energy could be one of the defining advancements of our time. As we explore and invest in various sustainable technologies, it is crucial not to overlook the significant potential of thorium reactors in shaping a cleaner, greener future.

Nuclear Waste Management and Minimization

In the dialogue about the future of nuclear power, one of the most daunting challenges is how to manage and minimize nuclear waste. This is not just a technical issue but also a social and political one, demanding attention and innovative solutions. As technology advances and the demand for cleaner energy sources grows, addressing the concerns related to nuclear waste becomes imperative.

Nuclear waste is a byproduct of nuclear reactors, fuel processing plants, and other facilities associated with nuclear processes. The primary concern with this waste is its radioactivity, which can persist for thousands of years. This long-lived hazard necessitates robust strategies for management and disposal, ensuring the safety of both current and future generations.

The first straightforward step in managing nuclear waste is categorization. Nuclear waste is typically divided into several categories based on its radioactivity level and longevity: high-level waste, intermediate-level waste, and low-level waste. Each category requires different handling, storage, and disposal methods, emphasizing the need for specialized approaches.

High-level waste (HLW), the most dangerous type, arises primarily from the spent fuel rods of nuclear reactors. Managing HLW involves cooling, containment, and eventual disposal in deep geological repositories. However, these repositories are not just large underground storage sites; they're meticulously engineered environments designed to isolate waste for thousands of years.

Several countries have made strides in this area. For instance, Finland is leading the way with its Onkalo deep geological repository, which is set to be operational in the near future. It's one of the first of its kind, serving as a model for other nations grappling with high-level waste management. The Onkalo project demonstrates that it's possible to create a secure, long-term solution for nuclear waste.

In parallel, intermediate-level waste (ILW) management involves solidifying the waste and storing it in secure facilities until a permanent disposal solution is available. ILW doesn't generate heat like HLW, but it still requires careful handling and robust containment strategies to prevent any environmental contamination.

Low-level waste (LLW), although less radioactive, still poses significant management challenges. This type includes items like clothing, tools, and other materials that have been exposed to radiation. Strategies for LLW involve thorough decontamination, compaction, and storage in shallow land burial sites, designed with multiple barriers to contain any potential leaks.

Another avenue for minimizing nuclear waste involves the advancement of nuclear reactor technology itself. Next-generation reactors, such as breeder reactors and thorium reactors, are designed not only to be more efficient but also to produce less waste. These technologies can utilize more of the nuclear fuel, thereby reducing the volume of spent fuel that becomes waste.

An innovative approach to waste minimization is reprocessing and recycling. In countries like France and Japan, reprocessing plants recover usable materials from spent nuclear fuel, which can then be used in new fuel assemblies. This process significantly reduces the volume of high-level waste and recovers valuable resources, making the entire nuclear fuel cycle more efficient.

Robotic technology also plays a crucial role in waste management. Robots can safely handle, sort, and package radioactive materials, minimizing human exposure to hazardous substances. Advances in artificial intelligence and robotics could further improve the accuracy and efficiency of these processes, presenting a promising frontier in nuclear waste management.

Public engagement and education are equally critical in successfully managing nuclear waste. Transparent communication about the safety measures, technological advancements, and long-term strategies is essential to gain public trust and support. Community involvement in decision-making processes can lead to more accepted and sustainable waste management solutions.

Regulatory frameworks must evolve to keep pace with technological advancements and societal expectations. International cooperation and harmonized regulations ensure that best practices are shared and that nuclear waste is managed uniformly and safely across borders. Organizations like the International Atomic Energy Agency (IAEA) play a pivotal role in setting global standards and providing guidance.

Funding and investment are fundamental to advancing nuclear waste management solutions. Governments, private sectors, and international bodies need to collaborate to ensure that necessary resources are allocated for research, development, and implementation of cutting-edge waste management technologies.

Addressing the issue of legacy waste, which includes the waste generated by early nuclear programs, is another significant challenge. Governments must prioritize the cleanup of these older sites, ensuring that past mistakes don't overshadow the benefits of current and future nuclear technologies. Remediation efforts at sites like Hanford in the United States or Sellafield in the UK showcase the magnitude and complexity of this task.

Finally, ethical considerations underpin the entire discourse on nuclear waste management. Ensuring intergenerational equity—whereby the responsibilities and risks of waste management are not unfairly transferred to future generations—is paramount. This ethical framework guides policies and practices, fostering a sustainable and responsible approach to nuclear energy.

In conclusion, the future of nuclear power hinges not just on advancing reactor technologies but also on effectively managing and minimizing nuclear waste. Through a combination of innovative technologies, regulatory frameworks, public engagement, and ethical considerations, it is possible to address the challenges of nuclear waste. By doing so, we can harness the full potential of nuclear power as a clean and sustainable energy source for the future.

Chapter 8:
The Promise of Fusion Energy

Fusion energy promises a revolutionary leap in how we can power our world, offering the tantalizing possibility of nearly limitless, clean energy derived from the same process that fuels the sun. Imagine harnessing this power on Earth; it's the epitome of sustainable energy solutions. Unlike traditional nuclear power that splits atoms (fission), fusion merges atomic nuclei, releasing vast amounts of energy and producing minimal long-lived radioactive waste. While the scientific community is still grappling with significant technological hurdles, advancements in materials science, particularly with high-performance superconductors, are steadily bringing us closer to viable fusion reactors. If we can crack this technological nut, fusion could fundamentally alter our energy landscape, providing a safe, reliable, and virtually inexhaustible energy source. This could be the cornerstone of a future where our energy demands are met without further harming our planet, making it a beacon of hope in our quest for sustainability.

Materials Challenges in Achieving Fusion

One of the most formidable barriers to achieving practical fusion energy lies in the materials that must endure the extreme conditions within a fusion reactor. Fusion reactions, akin to those powering the sun, produce temperatures exceeding millions of degrees Celsius. This necessitates the development of materials that can withstand intense heat, neutron bombardment, and radiation over prolonged periods. Traditional materials simply can't cut it under such conditions. Advanced alloys and ceramics, as well as novel composites, are under intense scrutiny to solve these daunting challenges. The process isn't just about surviving extreme environments; materials must also maintain structural integrity and function seamlessly for operational efficiency. Overcoming these obstacles will require breakthroughs that catalyze the next leaps in material science. As daunting as these challenges are, they hold the key to unlocking an inexhaustible, clean energy source, making the quest for fusion not just a scientific endeavor, but a crucial mission for our sustainable future.

The Role of Superconductors in Fusion Reactors is crucial to understanding how we can unlock almost limitless, clean energy from fusion reactions. Fusion, the process that powers the sun, has long been a dream of scientists and engineers because of its promise to provide a virtually inexhaustible energy source without the crippling drawbacks of fossil fuels or traditional nuclear power. Achieving this dream hinges on overcoming several daunting technological challenges, one of which is efficiently containing and controlling the extremely high temperatures necessary for fusion to occur. This is where superconductors come into play.

The primary role of superconductors in fusion reactors lies in their ability to create and maintain strong magnetic fields without losing energy to electrical resistance. In a fusion reactor, strong magnetic fields are required to contain the hot plasma—a state of matter made up of free electrons and nuclei—in which the fusion reactions occur. The extremely high temperatures, often exceeding 100 million degrees Celsius, make any physical containment vessel impractical. Instead, we use magnetic confinement to keep the plasma stable and prevent it from touching the reactor walls. Superconductors enable this magnetic confinement to be both powerful and sustainable over extended periods.

One of the most promising fusion reactor designs is the tokamak, a doughnut-shaped device that uses magnetic fields to stabilize and contain the plasma. Traditional materials for the magnetic coils in tokamaks have limitations due to their inherent electrical resistance, which leads to energy losses and the necessity for continuous power input to sustain the magnetic fields. However, when superconductors are used, they eliminate these losses by conducting electricity with zero resistance. This capability makes superconductors an indispensable component in the pursuit of practical and efficient fusion energy.

Superconductors work by entering a state of zero electrical resistance below a certain critical temperature, allowing them to conduct electricity indefinitely without losing energy. This property not only conserves energy but also drastically reduces the operational costs of sustaining the magnetic fields required for plasma containment. With operational superconducting magnets, we can generate the necessary strong and stable magnetic fields for longer durations, bringing us closer to achieving sustained nuclear fusion.

High-temperature superconductors (HTS) have been a game-changer for fusion research. Unlike traditional low-temperature superconductors, which require cooling to a few degrees above absolute zero, HTS materials operate at the much warmer temperatures of liquid nitrogen. This relative accessibility makes HTS more practical and cost-effective for large-scale applications like fusion reactors. The advent of HTS materials opens new possibilities for designing and constructing more efficient and powerful magnet systems for fusion reactors.

Historically, one of the biggest hurdles in fusion reactor development has been the enormous energy input required to start and maintain the reaction, compared to the energy produced.

Superconductors help to tip this balance by minimizing energy losses in the magnetic confinement system. For example, ITER, an international project aimed at demonstrating the feasibility of fusion power, relies heavily on superconducting magnets to achieve its objectives. The project's superconducting magnets are designed to produce magnetic fields over 11 Tesla in strength, significantly higher than what could be achieved with conventional materials.

Beyond just containment, superconductors also play a role in diagnostic tools within fusion reactors. By enabling more precise measurements and control, superconducting materials improve our ability to monitor and optimize the fusion process in real time. This means not only achieving a stable plasma state but also maximizing the efficiency and output of fusion reactions, bringing us closer to a commercially viable fusion power plant.

The environmental implications of integrating superconductors into fusion technology are profound. The potential for a virtually limitless and clean energy source would drastically cut our reliance on fossil fuels, reduce greenhouse gas emissions, and mitigate climate change. The materials required for superconductors, while not entirely devoid of environmental impact, typically involve less mining and processing than traditional energy sources. Furthermore, the efficiency and longevity of superconducting materials ensure that once they are in place, they require minimal ongoing energy input and maintenance.

Economic considerations also highlight the importance of superconductors in fusion reactors. While the upfront investment in superconducting technology is significant, the long-term savings from reduced energy losses and operational efficiencies make it an economically sound choice. Moreover, the development and deployment of superconducting technology in fusion reactors foster technological innovation and create high-tech jobs, contributing positively to the economy.

Another promising avenue involves investigating new superconducting materials that could operate at even higher temperatures or conditions optimized for fusion reactors. Research in this field is vibrant, with scientists constantly discovering new compounds and configurations that push the boundaries of superconductivity. These advancements could potentially simplify the cooling requirements and further reduce the costs and complexity of fusion reactors.

Furthermore, superconductors' role extends to the sustainability of energy systems. By integrating superconductors into the energy grid, we can enhance the storage and transmission of electricity generated from fusion reactors. Superconducting power lines, for instance, have the potential to transmit large amounts of electricity over long distances with minimal losses, ensuring that the energy produced is used as efficiently as possible.

Moreover, as we look to commercialize fusion energy, the scalability of superconducting technologies becomes critical. Superconductors can be tailored for various sizes and types of fusion reactors, from large-scale facilities like ITER to smaller, modular designs that could provide localized energy solutions. This flexibility enhances the adaptability and resilience of future energy systems, making them more robust against disruptions and capable of meeting diverse energy needs.

The integration of superconductors in fusion energy systems also ties into the broader narrative of sustainable resource use. By leveraging advanced materials like high-temperature superconductors, we not only make fusion more feasible but also drive innovation in material science and engineering. This can have a ripple effect, leading to the development of new materials and technologies that benefit other sectors and industries, further propelling us towards a sustainable future.

In conclusion, superconductors are not just a component of fusion reactors—they are a cornerstone of our journey towards realizing fusion power. Their unique properties address critical challenges in plasma containment, energy efficiency, and operational longevity. As we continue to innovate and refine these technologies, superconductors will undoubtedly play a pivotal role in making fusion energy a practical and sustainable reality. The promise of fusion energy, with superconductors at its heart, offers a vision of a future where clean, abundant energy is within our grasp, leading to profound environmental and economic benefits for generations to come.

Chapter 9:
The Circular Economy and
Resource Efficiency

In a world grappling with resource scarcity and environmental degradation, the principles of a circular economy offer a beacon of hope for a sustainable future. By designing products that prioritize durability and reusability from the get-go, we shift away from the traditional linear model of "take, make, dispose" and embrace a system that values the continual use of resources. This shift not only conserves materials but also significantly reduces environmental footprints. Innovation in recycling technologies plays a pivotal role here, transforming waste into valuable inputs for new products. Imagine a world where end-of-life products don't end up in landfills but are seamlessly reincorporated into the production cycle. This vision is not just possible—it's imperative. By maximizing resource efficiency and minimizing waste, we can pave the way for a more resilient and sustainable economy, ensuring that our technological advancements don't come at the cost of our planet's health.

Principles of Circular Economies

The essence of circular economies lies in redefining our production and consumption models to nurture sustainability. Traditional linear models follow a 'take, make, dispose' approach, depleting resources and generating waste. In stark contrast, circular economies aim to close the loop, ensuring that resources are reused, refurbished, and recycled indefinitely, mimicking natural ecosystems where nothing is wasted. Core principles include designing for longevity, modularity, and repairability, encouraging industries to adopt zero-waste philosophies, and embracing technologies that elevate recycling efficiencies. This paradigm shift not only conserves finite materials but also reduces environmental degradation, fostering a resilient and regenerative economic framework. As we stand on the brink of technological revolutions, integrating circular principles becomes imperative for sustaining the raw materials that will drive our future innovations. Thus, achieving a circular economy is not merely an option—it's a necessity to harmonize technological progress with ecological stewardship.

Redesigning Products for Durability and Reusability is not just a trend; it's a necessity in the quest for a sustainable future. The throwaway culture that dominated the 20th century has proven unsustainable, leading to an immense waste problem and the depletion of essential resources. For our technologies and lifestyles to be viable long-term, we must pivot towards products that not only last longer but can also be reused or repurposed multiple times before their life cycle ends.

First and foremost, the fundamental shift begins with *design philosophy*. Engineers and designers need to prioritize durability from the onset. This means selecting materials that boast long life spans, and structures that can withstand wear and tear. Consider the ubiquitous smartphone; instead of being designed for obsolescence within two years, what if it could last ten? Such a shift would not only reduce electronic waste but also conserve the precious metals that are often mined under controversial conditions.

Moreover, designing for reusability involves modularity. Imagine a washing machine where each component can be easily replaced if it fails. This would prevent the need for consumers to discard whole units due to a single broken part. Modularity makes repairs simpler and extends the product's life span. Europe is already seeing progress with its **Right to Repair** regulations, which aim to make electronic goods easier to fix. These regulations demonstrate how policy can influence design principles toward a more sustainable future.

Companies could adopt **circular design** principles that allow products to be broken down into their base materials at the end of their life cycle, ready to be turned into new items. For instance, certain sneaker brands are now crafting shoes that can be completely recycled. The materials used are selected with future disassembly in mind. These initiatives reduce waste and dramatically lessen the reliance on virgin materials.

Besides environmental benefits, longer-lasting products often offer economic advantages. Though the initial cost might be higher, the extended life span and reduced need for replacements can lead to lower long-term costs for consumers. This concept is evident in the rise of the refurbishment industry, where devices like laptops and smartphones are restored to near-new condition and sold at a fraction of the price of a new product.

Eco-friendly materials also play a crucial role in this paradigm. For example, some companies are now utilizing biodegradable plastics derived from plant materials. While these might not last as long as traditional plastics, they're designed to be part of a circular lifecycle, breaking down harmlessly when they're no longer useful. This is particularly vital for single-use items, where durability needs are lower, but reusability and recyclability are essential.

At the intersection of *durability and innovation*, we find initiatives like 3D printing. This technology allows for the creation of products with fewer weak points, potentially increasing their lifespan. Additionally, 3D printing often uses fewer resources and can utilize recycled materials, making it a potent tool for sustainable design.

We also mustn't overlook the value of **consumer education**. If people understand the long-term benefits of durable and reusable products, they're more likely to make sustainable choices. Educational campaigns and transparent information about product life cycles can inspire consumers to prioritize quality over quantity.

Retail models are transforming to match these shifts. Subscription services for items like furniture or electronics ensure that products are returned for refurbishment and reuse rather than being discarded. These services often include regular maintenance and upgrades, pushing the longevity of the products further. This model not only benefits the environment but also fosters a relationship between the consumer and the brand, as it aligns their interests towards sustainability.

Regulatory frameworks play an equally important role. Governments can incentivize manufacturers to create durable and reusable products through tax breaks, grants, and subsidies. Conversely, they can impose penalties for products designed with planned obsolescence. Properly structured regulation can drive the market towards innovation and sustainability.

While individual effort is crucial, it's clear that large-scale industry change is necessary. Companies with significant market influence have both the responsibility and capability to lead the charge. Apple's shift towards using recycled aluminum in its products sets a positive example. When industry leaders make these changes, they not only reduce their own environmental impact but also set a precedent for others to follow.

Collaboration across industries can amplify these efforts. For instance, partnerships between tech companies and recycling firms can create closed-loop systems where old products are systematically collected and repurposed into new ones. Such integrative efforts embed sustainability within the supply chain itself.

Technological advancements further bolster the move towards durability and reusability. Innovations in material science are yielding stronger, more versatile materials that can outlast their predecessors. Advanced polymers and alloys are just a few examples of the materials that hold promise for the next generation of durable goods.

To support these shifts, we must adopt a holistic perspective. It includes reimagining the entire lifecycle of products—from design and production to use and end-of-life management. The goal is to ensure that every step is as sustainable as possible, minimizing waste and maximizing value.

Ultimately, **Redesigning Products for Durability and Reusability** is essential if we are to meet the sustainability challenges of the future. It's not just about making things last longer; it's about creating a circular system where resources are perpetually reused, reducing our need to extract finite materials from the earth. By embracing these principles, we can pave the way for a more sustainable and resilient technological landscape.

Innovative Recycling Technologies

The circular economy thrives on a visionary perspective where waste doesn't exist. Instead, materials are perpetually cycled back into the economy, an ambition that's fueled by innovative recycling technologies. These advancements promise to transform discarded items into valuable resources, pushing the boundaries of sustainability like never before. From electronic waste to plastic, new technologies are redefining what's possible in the realm of recycling.

One of the frontrunners in innovative recycling is the advent of advanced sorting systems. Leveraging artificial intelligence and robotics, these systems can identify, sort, and process different types of waste with remarkable precision. Unlike traditional methods that rely heavily on manual labor, AI-driven sorters use sophisticated algorithms and machine learning to recognize materials and ensure they end up in the right recycling stream. This not only increases efficiency but also significantly reduces contamination rates, leading to higher-quality recycled products.

Equally groundbreaking is the development of chemical recycling. Unlike mechanical recycling, which can degrade the quality of materials like plastic, chemical recycling breaks down polymers into their original monomers. These monomers can then be repurposed to create new plastics of virgin quality. Technologies such as pyrolysis and depolymerization represent major strides in this field, offering solutions to problems that have long plagued conventional recycling processes.

Metal recovery from electronic waste is another area witnessing transformative innovation. Precious and rare metals found in electronics are incredibly valuable, yet challenging to recover. Traditional e-waste recycling methods often overlook smaller devices or recover these metals inefficiently. Recent advancements, however, involve hydrometallurgical processes and bioleaching, which utilize aqueous chemistry and microorganisms to extract metals in a more environmentally friendly way. These methods have the dual benefit of reducing the hazardous impact of e-waste on the environment while recovering substantial economic value.

Textile recycling, too, has seen significant advancements. The fashion industry, known for its environmental footprint, is embracing technologies that can break down fibers to their fundamental components, enabling the creation of new garments without virgin materials. This goes beyond simple reuse, as it involves a comprehensive deconstruction and rebuilding process. Innovations like enzyme-based recycling methods show great promise, providing a sustainable pathway for the fast fashion industry.

An inspiring example of innovation comes from the urban mining of construction and demolition waste. By effectively processing concrete, wood, and metals recovered from urban development projects, cities can drastically reduce landfill use and resource extraction. Advanced crushing, sorting, and purification technologies are leading this charge, enabling the recycling of materials that were once considered too complex or contaminated to reuse.

Glass recycling technologies have also evolved, making it possible to recycle glass indefinitely without loss of quality. Innovations in purification processes and the development of additives that improve the melting characteristics of recycled glass are expanding the possibilities of glass reuse across various industries. This stands in stark contrast to older methods that often relegated recycled glass to lower-value applications.

Moving on to organic waste, anaerobic digestion is changing the game by converting food waste and other organic matter into biogas and compost. This technology taps into naturally occurring microbes to break down waste in oxygen-free environments, generating renewable energy and nutrient-rich compost in the process. As a result, cities are now capable of turning organic waste—a significant part of the municipal waste stream—into valuable resources.

Plastic recycling has been particularly problematic due to the diversity of plastic types. However, solvent-based purification and the development of new catalysts that facilitate the breakdown of mixed plastics are providing more comprehensive recycling solutions. These innovations can handle complex mixtures of plastic types, something traditional mechanical recycling finds challenging. The goal is to create a closed-loop system where plastics can be perpetually recycled without quality loss.

A further frontier is the concept of upcycling, where waste materials are converted into products of higher value. This creative approach doesn't just recycle—it elevates. From turning discarded ocean plastic into high-quality footwear to transforming agricultural waste into bioplastics, upcycling takes what's old and makes it fundamentally new in a profoundly sustainable way.

Looking towards the future, the integration of blockchain technology in recycling processes promises enhanced transparency and efficiency. Blockchain can track the lifecycle of materials, ensuring that every step of the recycling process is documented and verified. This capability is particularly valuable for industries that require proof of sustainability and ethical sourcing within their supply chains.

The development of catalytic converters that can transform carbon dioxide emissions into useful products represents another visionary stride. By capturing CO_2 and converting it into fuels or industrial chemicals, these technologies turn a problem—greenhouse gases—into valuable inputs for other processes.

In addressing the issue of water-intensive recycling processes, innovations have emerged in the form of dry recycling systems that use minimal to no water. These systems are not only more sustainable but also reduce the overall operational costs of recycling facilities, showcasing yet another way in which technology can align economic and environmental goals.

Yet, as we stand on the cusp of these revolutionary changes, it is crucial to remember that technology alone can't carry the weight of a circular economy. Public policy, cross-sector collaboration, and consumer behavior must intersect harmoniously with these innovations. Only then can we fully harness the potential of these technologies to create a truly sustainable future.

The fascinating journey of Innovative Recycling Technologies teaches us that waste, as we know it, is nothing more than a resource in disguise. With continuous innovation, the materials of today's waste streams could very well become the raw materials of tomorrow's groundbreaking technologies.

Chapter 10:
Mining Innovations and Sustainability

As we delve deeper into the 21st century, the quest for sustainable resource management involves reimagining how we extract and utilize minerals essential for advancing technology. Reducing the environmental footprint of mining has grown increasingly urgent, demanding innovations that align with ecological preservation. Emerging green mining technologies, such as biomining and in-situ leaching, offer promising alternatives by minimizing landscape disruption and toxic by-products. Urban mining, the practice of reclaiming raw materials from discarded electronics and other waste, further encapsulates the forward-thinking approach required for a circular economy. These strategies not only replenish the supply chain with recycled resources but also diminish the need for new excavations, fostering a more sustainable and resilient future. By embracing these innovative practices, we can harmonize technological progress with the natural world, ensuring that we leave behind a greener planet for generations to come.

reducing the environmental impact of mining

Tackling the environmental challenges of mining is paramount if we are to pave the way for a truly sustainable future. Advanced technologies offer promising solutions, transforming traditional methods into more eco-friendly practices. For instance, innovations like precision mining, which utilizes data and automated machinery to minimize waste, and bioremediation, which harnesses biological processes to detoxify polluted environments, are beginning to change the landscape. Re-examining the supply chain and adopting lower-impact approaches can significantly reduce the ecological footprint of extraction activities. Additionally, policies promoting transparent environmental accountability encourage companies to invest in cleaner operations. By aligning technological progress with environmental stewardship, we can ensure that the raw materials essential for emerging technologies are sourced responsibly. This shift not only preserves ecosystems but also sets a precedent for sustainable industrial practices across various sectors.

Green Mining Technologies and Practices are transforming the landscape of resource extraction, ushering in an era where our need for vital materials doesn't come at the expense of the environment. By leveraging cutting-edge innovations, the industry is mitigating its impact on ecosystems while ensuring a steady supply of the critical elements that power our modern world.

Traditionally, mining has been synonymous with environmental degradation, involving deforestation, habitat destruction, and pollution. However, technological advancements are now enabling the sector to operate more sustainably. Energy-efficient machinery, precise extraction techniques, and improved waste management systems are just a few examples of how green mining is reshaping the industry's footprint.

At the heart of green mining lies the principle of reducing energy consumption. By employing automated and electrified machinery, mines can significantly decrease their reliance on fossil fuels. Robotics and artificial intelligence have also been pivotal, enhancing efficiency and reducing the human labor required, which in turn cuts down on the carbon emissions associated with traditional mining practices.

A key component of green mining is the development of advanced ore processing techniques that minimize waste and maximize yield. Hydrometallurgy, for example, uses aqueous chemistry to extract metals from ores, reducing the need for intense heat and energy. This not only conserves energy but also lowers the levels of harmful emissions like sulfur dioxide, which is common in conventional smelting processes.

Innovation in drilling technology has also reduced the environmental impact of mining. Directional drilling allows for more precise targeting of mineral deposits, meaning less disruption to the surrounding land. This technique is particularly beneficial for fragile ecosystems where traditional mining methods would be destructive.

Recycling water is another critical practice within green mining. By implementing closed-loop water systems, mines can reuse water multiple times, significantly reducing their overall water consumption. These systems often include advanced filtration technologies that ensure the water can be recycled without compromising its quality, thus protecting nearby waterways from contamination.

Furthermore, green mining isn't just about the extraction process but also about land rehabilitation post-extraction. Modern approaches ensure that once resources are depleted, the mined land is restored to its natural state or repurposed for other beneficial uses. This can include reforestation projects, the creation of wildlife habitats, or even the development of recreational areas for local communities.

Bioremediation is gaining traction as a sustainable method for dealing with mining by-products. By using microorganisms to break down and neutralize contaminants, mines can effectively clean up toxic waste without relying on chemicals that might cause further environmental harm. This technique is particularly useful for reducing heavy metals in the soil and water.

Equally important is the practice of tailings management. Tailings, the materials left over after the desired minerals have been extracted, often contain toxic elements. Innovative approaches like dry stacking, which involves removing the water from tailings and stacking the dry material, offer a safer and more sustainable way to handle these by-products. This minimizes the risk of tailings dam failures, which can have catastrophic environmental consequences.

The concept of green mining extends beyond Earth's surface. As we look towards the future, there's increasing interest in asteroid mining—a practice that, while still theoretical, promises to source critical materials without disrupting our planet's ecosystems. Space mining could serve as a pivotal component of the green mining framework, providing an even more sustainable solution for material extraction.

The industry is also exploring the potential of using renewable energy sources like solar and wind to power mining operations. By harnessing natural energy, mines can further reduce their carbon footprint and move away from reliance on non-renewable energy sources. This approach not only supports the sustainability goals of the mining sector but also sets a precedent for other industries to follow.

Collaboration and innovation are central to the successful implementation of green mining technologies. Partnerships between academia, industry, and governments are fostering research into new methods and technologies, while regulatory frameworks are increasingly supporting sustainable practices. This collective effort ensures that the advancements in green mining continue to evolve and improve.

Economic incentives are also playing a crucial role in driving the adoption of green mining technologies. Governments and organizations are offering grants, tax breaks, and other financial benefits to companies that demonstrate a commitment to sustainable practices. These incentives not only encourage more companies to innovate but also ensure that green mining becomes the standard rather than the exception.

Public awareness and consumer demand are additional factors pushing the mining industry towards greener practices. As more people become aware of the environmental impacts of resource extraction, there's a growing demand for products sourced through sustainable means. This consumer pressure is prompting companies to adopt greener practices to maintain their market position and reputation.

Finally, the future of green mining looks promising as we continue to innovate and adapt to new technologies and methodologies. By continuously improving and refining these practices, we can hope to achieve a balance where the need for vital resources doesn't come at the expense of our planet's health. The journey towards sustainable mining is ongoing, but the steps being taken today are paving the way for a more sustainable and efficient future.

Urban Mining: A New Source of Materials

Urban mining is emerging as a transformative approach to sourcing materials by tapping into the wealth of resources embedded in our cities. Unlike traditional mining, which involves extracting raw materials from the earth's crust, urban mining focuses on recovering valuable metals and other components from electronic waste and other urban debris. This approach not only addresses the growing issue of electronic waste but also contributes significantly to a circular economy.

The concept of urban mining is rooted in the recognition that cities are rich repositories of discarded electronics, buildings, and infrastructure, all of which contain valuable materials. As urban areas continue to expand and technology advances, the volume of electronic waste, or e-waste, has skyrocketed. According to the Global E-waste Monitor, the world generated a record 53.6 million metric tons of e-waste in 2019, a figure projected to increase in the coming years.

One of the largest sources of materials suitable for urban mining is electronic waste. This waste includes discarded smartphones, computers, televisions, and other electronic devices. These products are often rich in precious metals like gold, silver, and platinum, as well as rare earth elements essential for modern technology. For example, one ton of e-waste has more gold in it than 17 tons of gold ore.

Urban mining has significant environmental benefits. By reclaiming metals and other materials from e-waste, we can reduce the demand for primary mining and the associated environmental toll. Traditional mining operations are notorious for their environmental impact, from deforestation and habitat destruction to water pollution and carbon emissions. In contrast, urban mining can be conducted within existing urban infrastructures, minimizing land disruption and reducing greenhouse gas emissions.

Another compelling aspect of urban mining is its potential to create economic opportunities. As the e-waste recycling industry grows, it creates jobs in collection, sorting, dismantling, and processing. Moreover, urban mining can drive technological innovation in recycling methods, making the process more efficient and cost-effective. Companies are developing advanced techniques to extract materials from e-waste, including hydrometallurgical processes and bioleaching, which use microbes to break down metals.

While the concept of urban mining is promising, it also faces several challenges. One major hurdle is the need for effective e-waste collection and sorting systems. Many countries still lack the infrastructure to manage e-waste properly, leading to significant volumes being sent to landfills or incinerated. This not only wastes valuable resources but also poses environmental and health risks due to the toxic elements often found in electronic devices.

Implementing urban mining on a large scale requires collaboration between governments, industry, and the public. Governments can play a pivotal role by enacting policies that promote e-waste recycling and urban mining. Extended producer responsibility (EPR) laws, for instance, require manufacturers to take back their products at the end of their lifecycle, ensuring they are recycled properly. Public awareness campaigns can also educate consumers about the importance of recycling e-waste and the benefits of urban mining.

From an industry perspective, design-for-recycling principles can make urban mining more effective. Manufacturers can develop products that are easier to disassemble and recycle, using materials that can be readily reclaimed. This not only facilitates urban mining but also aligns with the broader goals of a circular economy, where products are designed for longer lifespans and greater recyclability.

Looking ahead, advancements in technology will continue to enhance the efficiency and profitability of urban mining. Artificial intelligence (AI) and robotics can be used to automate the sorting and dismantling of e-waste, improving accuracy and reducing labor costs. Machine learning algorithms can also optimize processing methods, identifying the most efficient ways to recover valuable materials from complex mixtures.

Another promising avenue for urban mining is the recovery of materials from building and infrastructure demolition. Modern buildings are constructed with a variety of materials, including metals, plastics, and composites, many of which can be recycled. Urban mine sites often lie in abandoned or underutilized areas, presenting an opportunity to salvage materials while revitalizing urban landscapes.

Resource scarcity will likely drive the expansion of urban mining. As traditional mines are depleted and environmental regulations tighten, the cost and feasibility of sourcing materials from the earth will continue to rise. Urban mining offers a sustainable alternative, harnessing the resources already present in our urban environments. This shift could reduce the global reliance on primary mining, contributing to more resilient and sustainable supply chains.

Urban mining is also closely linked to the broader principles of the circular economy. By recovering and reusing materials from products at the end of their life cycle, we can create closed-loop systems that minimize waste and resource consumption. This not only conserves natural resources but also reduces the environmental footprint of production and consumption.

In conclusion, urban mining represents a new frontier in sustainable resource management. As cities grow and technology evolves, the potential to reclaim valuable materials from urban environments will only increase. By embracing urban mining, we can reduce the environmental impact of traditional mining, create economic opportunities, and move closer to a circular economy. It's an opportunity to turn waste into wealth, transforming our urban landscapes into resource-rich environments that support a more sustainable and innovative future.

Urban mining isn't just a novel approach to recycling; it's a critical strategy for achieving sustainability in an increasingly resource-constrained world. Through continued innovation, policy support, and public engagement, we can unlock the full potential of urban mining, securing the materials we need for the technologies of tomorrow while protecting our planet for future generations.

As we continue to explore and innovate in the realm of urban mining, it's essential to recognize its broader implications for the global resource landscape. Developing scalable and efficient urban mining processes will be a key component of our transition to a more sustainable and circular economy. By turning our cities into urban mines, we're not just addressing the challenges of e-waste and resource scarcity; we're pioneering a new paradigm of resource stewardship and environmental responsibility.

Chapter 11:
Alternative Materials:
Beyond the Conventional

As we continue to push the boundaries of technology and sustainability, exploring alternative materials becomes imperative. These aren't just mere substitutes for existing materials but groundbreaking innovations that promise to revolutionize entire industries. Materials like graphene and nanomaterials are at the forefront, representing leaps in strength, conductivity, and versatility. Imagine the potential when these materials are harnessed for electronics, energy storage, and even medical advancements. Beyond their extraordinary properties, they carry the promise of more sustainable and efficient production chains. This chapter delves into these wonder materials, assessing their capabilities and the profound implications they have on our future technologies. By leveraging these advanced alternatives, we open new pathways to sustainability, driving forward the dream of a more environmentally responsible and technologically advanced world.

Graphene: The Miracle Material

When we talk about pushing the boundaries of technology with sustainable and revolutionary resources, graphene stands out as a game-changer. Just a single layer of carbon atoms arranged in a hexagonal lattice, graphene boasts remarkable attributes—it's incredibly strong yet lightweight, flexible, and an excellent conductor of electricity and heat. Imagine a material that's 200 times stronger than steel but so thin that a single gram could cover a football field. Its potential seems almost limitless, from supercharging electronic devices to creating ultra-efficient solar panels and even revolutionizing medical treatments. But the allure of graphene goes beyond its physical properties; it also holds promise for sustainable development. As we strive to reduce our carbon footprint, graphene's use in lightweight composites for transportation can lower energy consumption, and its role in advanced battery technologies might pave the way for greener energy storage solutions. By embracing graphene, we aren't just witnessing the dawn of a new technological era; we're also taking significant strides toward a sustainable and innovative future.

Applications in Electronics and Beyond The story of graphene, often touted as the "miracle material," continues to unfold with astonishing potential, particularly in the realm of electronics. Renowned for its unique combination of strength, flexibility, and conductivity, graphene has the ability to transform industries far beyond our imagination. While its initial appeal was associated with the electronics sector, ongoing research illustrates much broader applications.

Inside the electronics industry, graphene has already begun to shake up conventional paradigms. Known for its exceptional electrical properties, graphene enables the development of faster and more energy-efficient transistors. For example, graphene transistors can potentially outperform silicon transistors, driving the advent of more powerful and compact computers and mobile devices. This advancement is not limited to enhancing processing speed; it also offers a reduction in power consumption, which aligns well with global energy sustainability goals.

Graphene's application extends to the realm of flexible electronics. Imagine bendable smartphones or rollable tablets that maintain high performance and durability. Such innovations could revolutionize how we interact with technology, making portable electronics more rugged and adaptable. With graphene, the dream of flexible, wearable tech gears becomes feasible, providing smart textiles and wearable gadgets with heightened sensitivity and resilience.

Beyond traditional electronics, graphene's properties are poised to revolutionize energy storage systems, particularly batteries and supercapacitors. Modern-day batteries, despite their advancements, are limited by energy density and recharge times. Here, graphene steps in to potentially solve these issues. Graphene-enhanced batteries promise rapid charge and discharge cycles while improving energy storage capacity. This will support the growing demand for electric vehicles and portable power devices, accelerating the transition to a more energy-efficient future.

In addition to batteries, graphene's role in supercapacitors is equally transformative. Supercapacitors, known for their ability to charge rapidly and provide bursts of energy, can benefit immensely from the inclusion of graphene. The material's high surface area and conductivity can significantly enhance the performance of supercapacitors, making them a viable alternative or supplement to traditional batteries in various applications. Consequently, this could lead to advances in many sectors, including transportation, consumer electronics, and renewable energy storage.

The influence of graphene reaches even further into complex realms like quantum computing. Graphene's unique electronic properties make it a promising material for the construction of qubits – the fundamental units of quantum information. Unlike classical bits, qubits can exist in multiple states simultaneously, offering an exponential increase in computing power. As research progresses, graphene might become instrumental in realizing the long-sought-after quantum computing revolution.

On the environmental front, graphene holds potential beyond its electrical properties. Its remarkable filtration capabilities are promising for water purification systems, enabling the removal of contaminants and salts at unprecedented efficiency. This application is particularly vital as we face growing global water scarcity issues. Incorporating graphene in desalination and filtration systems could help secure clean water sources, crucial for both human consumption and agricultural use.

In the realm of composites, graphene can dramatically enhance material performance. By integrating graphene into composites, we can develop materials that are stronger, lighter, and more resilient than their traditional counterparts. This has profound implications for aerospace, automotive, and construction industries. Imagine airplanes that weigh less yet are more robust or car bodies that are more resistant to impact. The potential to revolutionize these sectors cannot be understated.

Moreover, graphene's thermal conductivity extends its applications into heat management solutions. As electronic devices become more powerful, managing heat dissipation becomes a critical concern. Graphene can be used in thermal interface materials, facilitating efficient heat removal from electronic components, thus enhancing performance and lifespan. This is particularly important in high-performance computing and LED technologies, where efficient cooling is paramount.

Healthcare is another field ripe for graphene's influence. Already, we are seeing the use of graphene in biosensors capable of detecting diseases at an early stage. These sensors are incredibly sensitive and can be used for a variety of diagnostic applications, from glucose monitoring to detecting pathogens. The implications for early detection and treatment of diseases are profound, potentially transforming how healthcare is delivered and improving patient outcomes significantly.

In addition to biosensors, graphene's biocompatibility opens new frontiers in medical devices and drug delivery systems. Imagine a world where implants are not only more durable but also capable of delivering medication directly to targeted areas within the body. The use of graphene in medical devices could lead to significant advancements in surgical implants, tissue engineering, and targeted drug delivery, improving patient care and treatment efficiency.

The possibilities of graphene even stretch into the environmental sector through the realm of air purification. Its ability to adsorb pollutants makes it an excellent candidate for air filters that can trap particulate matter and harmful gases. As we face growing concerns over air quality and pollution, graphene-filter technologies could provide a much-needed solution, improving the air we breathe in both urban and industrial settings.

Graphene's reach doesn't stop at earthbound applications. Space exploration stands to gain significantly from its properties, particularly in the development of lightweight, resilient materials for spacecraft. Reducing the weight of space equipment while enhancing its durability is crucial for long-term space missions. With graphene-enhanced materials, we can build spacecraft that are not only more efficient but also have extended operational lifespans, potentially reducing the cost and increasing the feasibility of future space endeavors.

As we delve deeper into the 21st century, the applications of graphene continue to expand, touching various facets of technology and industry. The material's potential to drive innovation while promoting sustainability makes it a cornerstone of future technological advancements. However, the challenge lies in scaling up production and making graphene-based technologies accessible and affordable.

Being ambitious means tackling these challenges head-on, inching us closer to a world where the full potential of graphene is realized, and its benefits are universally felt.

In summary, graphene's transformative potential in electronics and beyond epitomizes the kind of innovation that can drive us toward a sustainable and technologically advanced future. Exploring its applications offers a glimpse into a world where technology and sustainability are not just compatible but symbiotic. This journey of harnessing graphene's potential will require collaborative efforts across industries and disciplines, but the rewards—efficiency, innovation, and sustainability—are well worth the pursuit.

Nanomaterials and Their Role in Future Technologies

As we continue to push the boundaries of engineering and science, nanomaterials are stepping into the spotlight as game-changers for future technologies. These materials, measured in the scale of nanometers, possess unique physical and chemical properties that can be harnessed in ways that were previously unimaginable.

Nanomaterials, characterized by their extremely small size and high surface area to volume ratio, offer a landscape teeming with new possibilities. For instance, imagine electronics becoming smaller and more efficient or medical treatments being delivered with pinpoint precision. These are not just abstract ideas but are tangible realities that nanomaterials can bring to life.

Let's consider their application in computing. Traditional silicon-based chips are reaching their physical limitations. As we demand faster and more powerful computing, the miniaturization plateau becomes a significant challenge. Nanomaterials, such as carbon nanotubes and quantum dots, are poised to revolutionize this field, promising to surpass the performance of conventional semiconductor materials.

Furthermore, the role of nanomaterials in sustainability extends beyond enhancing current technologies. In the realm of energy, for example, nanomaterials can significantly improve the efficiency of solar panels and the capabilities of batteries. Quantum dot solar cells and nanoscale battery materials can radically transform our approach to harnessing and storing energy, making renewable resources much more viable.

Take the case of quantum dot solar cells. These cells can be tuned to absorb different parts of the solar spectrum by simply adjusting their size. This ability to tailor their properties means a higher efficiency in converting sunlight into electricity, surpassing the traditional silicon-based solar panels most people are familiar with.

Batteries, another critical area where nanomaterials can make a difference, benefit from the high surface area of nanostructured materials. Imagine a battery that charges in minutes instead of hours, or one that has a significantly longer lifespan. Materials like silicon nanowires or graphene can achieve these improvements by facilitating faster electron and ion transport, which is crucial for battery performance.

Of course, the medical field is not left out of this transformative potential. Drug delivery mechanisms that employ nanomaterials can target specific cells without affecting the surrounding healthy tissue. This precision not only improves the efficacy of treatments but also minimizes side effects, a crucial advancement in cancer therapies and other chronic diseases.

The potential of nanomaterials in creating more sustainable fabrics and textiles is another exciting frontier. By incorporating nanomaterials, fabrics can become more durable, stain-resistant, and even capable of self-repair. Imagine clothing that adjusts to temperature changes, insulates better, and lasts longer, thereby reducing waste and consumption.

Environmental cleanup also finds a powerful ally in nanomaterials. Nanoparticles can be used to efficiently detect and remove contaminants from water and soil. This could revolutionize the way we tackle pollution, making environmental remediation faster and more effective. For instance, nanomaterials like titanium dioxide are being utilized to break down pollutants in water through photocatalytic reactions, providing cleaner water supplies with minimal energy input.

However, the adoption of nanomaterials isn't without challenges. Addressing safety and ethical concerns is paramount. The unique properties that make nanomaterials beneficial also pose unknown risks. Their small size might enable them to penetrate biological barriers or accumulate in the environment in unforeseen ways. Rigorous testing and regulatory frameworks are necessary to ensure that these new materials do not inadvertently create new problems.

International cooperation and policy development are crucial in regulating and guiding the use of nanomaterials. Transparent guidelines need to be established to address any potential health and environmental impacts. Only through concerted global efforts can we harness the full potential of nanomaterials while safeguarding against their risks.

The economic implications of nanomaterials also merit consideration. Investing in the development and mass production of these materials could trigger significant shifts in market dynamics. Emerging economies could find new opportunities in this high-tech field, altering the landscape of global trade and industrial dominance.

Education and workforce development are essential in preparing for the nanomaterials revolution. Incorporating comprehensive nanotechnology programs in universities and training centers will ensure that the next generation of scientists, engineers, and technicians are equipped with the necessary skills to drive this field forward.

Moreover, public awareness and engagement play vital roles. As consumers and citizens, understanding the implications of nano-technology can foster informed decision-making and support for regulations and innovations that prioritize sustainability. Encouraging curiosity and knowledge about nanomaterials can lead to a society that is both innovative and conscious of the ethical dimensions of technological progress.

In conclusion, nanomaterials stand at the cusp of transforming multiple industries, heralding a future where technology is more efficient, sustainable, and responsive to human needs. By addressing the inherent challenges and embracing the opportunities they present, we pave the way for a world that leverages the extraordinary potential of the smallest components of matter to create monumental changes for the better.

The journey toward integrating nanomaterials into mainstream applications is still unfolding. With ongoing research, cross-disciplinary collaboration, and a steadfast commitment to sustainable practices, these tiny materials could indeed hold the key to unlocking a future of endless possibilities.

Chapter 12:
Water: Sustaining the Source of Life

As we move forward in our exploration of critical resources, it's crucial to pause and consider water, the lifeblood of our planet and the key to sustaining future technologies and societies. The importance of water can't be overstated—it underpins agricultural production, industrial processes, energy generation, and even our own daily lives. Advances in water purification technologies are making headlines, but just as vital are the everyday conservation techniques that can save millions of gallons. We'll delve into how innovative approaches like nanofiltration and biomimicry are revolutionizing water treatment. Concurrently, the adoption of efficient irrigation systems and smart water management practices are critical in making every drop count. The growing pressures of climate change and population growth make water stewardship an imperative for both immediate human needs and long-term sustainability. Let's get inspired to innovate and protect this irreplaceable resource, as the way we manage water today will define the future resilience of our ecosystems and technologies.

Advancements in Water Purification Technologies

As we delve into the ongoing advancements in water purification technologies, it's clear that an innovative and sustainable approach to managing this vital resource is not just possible but essential. Emerging techniques are leveraging nanotechnology, advanced filtration systems, and even biological processes to make water purification more efficient and less environmentally taxing. Technologies like graphene-based filters promise to revolutionize the field with their exceptional ability to remove contaminants at a molecular level. Meanwhile, biodegradable filtration materials are being developed to reduce waste. These solutions don't just mitigate water scarcity; they also offer a blueprint for sustainable resource management. This progress is not just about technological capabilities, but about envisioning a future where clean water is accessible to all, underscoring our collective responsibility to foster a healthier planet.

Making Every Drop Count: Water Conservation Techniques stands as a crucible for the future of sustainable living. Given that water is a finite and vital resource, the imperative for conservation has never been more pressing. Water conservation isn't only about reducing consumption—it's also about adopting innovative techniques and technologies designed to maximize efficiency across various sectors. Whether in agriculture, industry, or our homes, every drop truly counts.

First, let's look at agriculture, which typically consumes about 70% of global freshwater. Employing more efficient irrigation methods, such as drip or micro-sprinkler systems, can drastically reduce water waste. Traditional flood irrigation methods allow much of the water to evaporate or runoff before it even reaches the crops. Precision agriculture, aided by data analytics and sensors, can optimize water application based on real-time plant requirements, promising significant water savings.

Industries, particularly those involved in manufacturing and energy production, are notable water consumers. Innovations such as closed-loop systems, where water is continuously recycled within the plant, reduce fresh water intake. Techniques like membrane bioreactor (MBR) technology help in treating wastewater to a quality suitable for reuse, closing the loop on industrial water use. Moreover, process efficiency improvements can reduce the water footprint by requiring less water per unit of product.

Residential water use, while smaller in comparison, presents ample opportunities for conservation. Low-flow fixtures in showers, faucets, and toilets can reduce water consumption dramatically. Rainwater harvesting systems allow households to capture and store rainwater for non-potable uses, such as irrigation and toilet flushing. Greywater systems, which recycle water from sinks, showers, and washing machines, offer another layer of conservation by reusing water that would otherwise go down the drain.

In urban planning, designing green spaces with drought-resistant native plants, known as xeriscaping, decreases the need for watering. Smart irrigation controllers, which adjust watering schedules based on weather forecasts, soil moisture levels, and plant needs, can also make urban landscapes more water-efficient. Urban planners are increasingly looking at permeable pavements and green rooftops to enhance water capture and reduce runoff, promoting a holistic approach to urban water conservation.

Another promising frontier is the use of IoT (Internet of Things) devices for water management. IoT sensors can monitor flow rates, detect leaks, and even predict system failures before they occur, thus preventing water loss. Smart metering systems empower consumers to track their water usage in real-time, fostering more conscious consumption habits.

Behavioral changes play a critical role in conservation efforts. Educating the public on water-saving practices, such as taking shorter showers, turning off the tap while brushing teeth, and using dishwashers only with full loads, can collectively make a substantial difference. Awareness campaigns that communicate the baffling stat that a leaking faucet can waste over 3,000 gallons of water a year can spur immediate and impactful action.

At the policy level, governments can incentivize water-saving technologies and practices through subsidies and tax breaks. Regulations enforcing the installation of water-efficient appliances in new buildings can be a significant step forward. Water pricing models that reflect the true cost of water encourage conservation and investment in water-saving technology.

Water conservation extends to protecting and revitalizing natural water bodies. Wetland restoration projects help in water purification, flood control, and groundwater recharge. Preserving river ecosystems ensures the sustenance of both biodiversity and water supplies for human use. Strategic tree planting along watercourses not only stabilizes banks but also helps in maintaining the hydrological cycle.

Research and development in advanced water treatment technologies promise to bolster our conservation efforts. Innovations like forward osmosis and capacitive deionization are emerging as energy-efficient alternatives to traditional desalination processes. Such technologies can make previously unusable water sources a viable option, reducing the burden on freshwater reservoirs.

International cooperation is critical in addressing the global challenge of water scarcity. Shared water resources can be a source of conflict or cooperation, depending on how they are managed. Multinational agreements and partnerships can promote best practices, technology transfers, and coordinated conservation efforts, ensuring equitable access to water across borders.

In considering the economic aspects, investing in water-saving technologies can result in long-term financial benefits. Reduced operational costs, lower utility bills, and compliance with regulatory standards can provide a competitive edge. Moreover, demonstrating a commitment to sustainability can enhance an organization's reputation and brand value, making water conservation not just an environmental imperative but also a strategic business decision.

The cultural and social dimensions of water conservation should not be underestimated. Indigenous knowledge and traditional water management practices offer a trove of sustainable techniques. Engaging local communities in water conservation efforts ensures that solutions are contextually appropriate and widely accepted. Social norms evolve when conservation becomes a shared value, leading to widespread behavioral change.

Ultimately, making every drop count means recognizing water for what it is—a finite, invaluable resource that sustains life and drives technological progress. As we look to the future, our efforts to conserve water must be as diverse and innovative as the challenges we face. By integrating technological advances with individual and collective actions, we can build a sustainable water future that supports both human prosperity and ecological integrity.

In closing, we move forward with the understanding that water conservation is a multifaceted challenge requiring a concerted effort across disciplines and sectors. The journey to making every drop count is both a technological and societal endeavor, one that demands our relentless commitment and ingenuity.

Chapter 13:
Air: The Quest for Quality

As we forge ahead into the future, the air we breathe becomes not just a matter of health, but of survival and innovation. We've already seen the devastating effects of pollution on our ecosystems and personal well-being. The challenge now lies in not just mitigating these effects but reversing them through groundbreaking advancements in air purification. Pioneering technologies and advanced materials hold the promise of cleaner air, reducing the toxic load on our lungs and planet. This quest for quality air intertwines with all aspects of modern tech and sustainable practices, steering us toward a future where quality air isn't a luxury but a fundamental right. Our journey involves transforming industrial emissions, combating urban smog, and creating environments in which cleaner air contributes to healthier communities and a more resilient planet.

Innovative Solutions for Air Purification

As global pollution levels reach unprecedented heights, innovative solutions for air purification have become critical to ensuring a sustainable, healthy future. Cutting-edge technologies such as nanomaterials, advanced filtration systems, and bioreactors are stepping up to tackle airborne pollutants more effectively than ever before. Imagine air-cleaning devices embedded within urban infrastructure, silently scrubbing the air of harmful particulates and gases as you walk down the street. Additionally, advancements in material science are paving the way for smart fabrics that can capture and neutralize contaminants directly from the air we breathe. Even traditional methods are getting a tech upgrade: forests and green spaces, nature's air purifiers, can now be augmented with IoT sensors to monitor and optimize their air-cleaning capabilities. These innovations not only promise cleaner air but also inspire a vision of cities that actively contribute to the health of their residents. Transformative ideas like these make it clear that the quest for quality air is not just an environmental challenge; it's a call to reimagine our future with creativity and a commitment to sustainability.

Combating Pollution Through Advanced Materials is more than just a technological aspiration; it represents a shift in how we think about and interact with our environment. The issue of air pollution dates back centuries, yet our modern industrial lifestyle exacerbates it like never before. Fortunately, cutting-edge materials technology offers promising solutions for this pervasive challenge.

The air around us is contaminated with a mix of pollutants - from particulate matter to greenhouse gases such as carbon dioxide (CO_2) and methane (CH_4). These pollutants not only contribute to climate change but also pose severe health risks. Recent advancements in material science could lead us to breathable cities and a genuinely cleaner atmosphere.

Among the standout innovations is the development of advanced filtration materials. Traditional air filters, often constructed from activated carbon and synthetic fibers, are effective but come with limitations, including disposal issues and limited filtration capacity. Enter the era of nanomaterials, whose unique properties can trap pollutants more efficiently and withstand longer usage, reducing the frequency of replacements.

Nano-coated surfaces are one key advancement in this area. These surfaces leverage materials like titanium dioxide (TiO_2) that possess photocatalytic properties. When exposed to sunlight or ultraviolet light, these surfaces break down hazardous pollutants into less harmful substances. This transformative process could be applied to outdoor facades and public infrastructure, making our cities cleaner without the need for complex machinery.

Another fascinating development is the utilization of metal-organic frameworks (MOFs). MOFs are crystalline materials composed of metal ions and organic molecules that form porous structures. Their customizable architecture allows them to trap specific pollutants with exceptional efficiency. Imagine buildings cloaked in MOF-based coatings, absorbing CO_2 and NO_x, contributing to air purification while passively enhancing urban aesthetics.

Graphene, often heralded as the "wonder material," shows promise in air purification as well. Its high surface area and conductivity make it an effective medium for adsorbing pollutants. Researchers are exploring the potential of graphene-based membranes for capturing fine particulates and even filtering gases. These membranes, lightweight and highly efficient, could revolutionize mobile air purification devices.

The utility of advanced materials extends beyond passive solutions. Active technologies, such as electrochemical cells incorporating these materials, can transform pollutants into useful products. For instance, converting CO_2 into hydrocarbon fuels or other chemicals not only cleanses the air but also adds economic value, creating a cyclical benefit.

Let's also consider the role of biomimetic materials in our fight against air pollution. The emulation of natural processes, such as the way leaves photosynthesize, can inspire innovative materials that capture and convert pollutants. These biologically inspired materials can enhance urban green spaces, turning parks and rooftops into active environmental cleaners.

Not to be overlooked is the economic and societal impact. Implementation of these advanced materials in air pollution control is not merely a technical challenge but a socio-economic endeavor. Governments and industries must recognize the long-term benefits and be willing to invest in large-scale deployments and infrastructure upgrades. The societal shift toward accepting and supporting these innovations is crucial for their success.

Education and public awareness programs can bridge the gap between technological innovation and community acceptance. By informing citizens about the benefits of advanced materials and their potential impact on air quality, we can foster a supportive environment for these technologies. Public engagement can accelerate adoption and lead to policies that endorse further research and deployment.

The battle against air pollution through advanced materials isn't just about cleaner air; it's about transforming our relationship with the environment. These materials present opportunities to rethink urban planning, architectural design, and public health strategies. As these technologies develop, they'll become integral to creating sustainable and resilient communities.

A future where skyscrapers purify the air and public parks act as urban lungs is not just an idealistic vision but a tangible goal. The frontiers of material science are pushing boundaries, and the integration of these advancements into our daily lives could lead to unprecedented improvements in air quality and overall ecological health.

Given the risks associated with air pollution, including its ties to respiratory illnesses and climate change, the urgency to deploy advanced materials is paramount. Yet, the journey ahead requires coordinated efforts from scientists, entrepreneurs, policymakers, and citizens alike. By harnessing the full potential of these innovations, we can inhibit the sources of pollution while fostering a cleaner, healthier environment.

In conclusion, "Combating Pollution Through Advanced Materials" not only encapsulates a series of scientific pursuits but a broader vision of societal transformation. This vision encompasses technological innovation, economic foresight, and a collective will to improve our planet's health. The raw materials of the future, in this context, are those that enable sustainability and enhance our quality of life.

Chapter 14:
Soil: The Foundation of Our Future

Soil is often referred to as the "skin of the Earth," and for good reason. It's the fundamental layer that supports not just plant life, but entire ecosystems and agriculture crucial for human survival. As we're venturing forward in technology and sustainability, addressing the condition of our soil couldn't be more critical. The state of our soil directly impacts crop yields, water filtration, and even carbon sequestration. Implementing cutting-edge soil remediation technologies can restore degraded lands, making them arable again. Moreover, by utilizing innovative materials, we can promote sustainable agriculture. Imagine a future where nutrient-rich soils enhance food security and minimize our dependence on chemical fertilizers. The health of our soil today will determine the resilience of our food systems tomorrow. Taking action now paves the way for a greener, more sustainable planet, ensuring that future technologies are rooted in strong, fertile grounds.

Soil Remediation Technologies

In our quest for a sustainable future, soil remediation technologies are emerging as pivotal tools. These innovative methods address the contamination and degradation of soil, reinstating it as a vital resource for ecosystems and agriculture alike. From phytoremediation, which leverages plants to absorb and break down pollutants, to advanced chemical treatments and bioremediation practices utilizing microbes, each approach contributes uniquely to restoring soil health. A crucial aspect of these technologies is their potential to transform wastelands into productive terrains, enabling resilient agricultural systems and reducing the ecological footprint of industrial activities. As we forge ahead, ensuring soil remains fertile and uncontaminated becomes not merely a matter of environmental stewardship but a cornerstone of our collective aspiration towards a sustainable and technologically advanced society.

Promoting Sustainable Agriculture Through Innovative Materials is more than just a catchphrase; it is a necessary revolution in how we approach feeding our future. Today's agricultural systems face monumental challenges, including soil degradation, climate change, and increasing global food demand. It's clear that the traditional methods of agriculture are no longer sufficient. That's where innovative materials come into play, offering transformative solutions to create more resilient and sustainable agricultural practices.

First, let's delve into the importance of soil health. Healthy soil is the cornerstone of sustainable agriculture, providing essential nutrients to crops and acting as a natural water filter. However, many conventional farming practices degrade soil quality. Innovations like biochar—a type of charcoal that improves soil fertility—offer a beacon of hope. Biochar not only enhances nutrients in the soil but also sequesters carbon, thereby mitigating climate change.

Biochar's benefits don't stop there. One of its most promising features is its long-term stability. Unlike typical organic matter, biochar is resistant to decomposition, allowing it to remain in the soil for hundreds of years. This longevity makes it a sustainable method for improving soil health. Researchers are also developing various biochar-based products tailored to specific soil types and crops, making it a versatile solution for diverse agricultural needs.

Nanotechnology also holds incredible potential for sustainable agriculture. Nanomaterials can be designed to improve plant growth, nutrient use efficiency, and resistance to pests and diseases. Nanofertilizers, for instance, offer controlled-release mechanisms that supply nutrients precisely when and where plants need them. This not only leads to higher crop yields but also minimizes the environmental impact of excess nutrient runoff that contaminates water bodies.

Another promising innovation is the use of biodegradable materials in farming practices. Traditional plastics, commonly used in mulch films and packaging, contribute to environmental pollution. Replacing these with biodegradable alternatives made from plant-based materials can dramatically reduce waste. Biodegradable mulch films, for example, not only protect crops but also break down into organic matter, enriching the soil.

Water is another critical resource in agriculture, and innovative materials are changing the game in irrigation efficiency. Hydrogel-based materials, designed to hold water and release it slowly to plant roots, can significantly reduce water consumption. These super absorbent polymers expand upon absorbing water and slowly release moisture, ensuring plants receive a steady supply even during dry spells. This technology is particularly valuable in regions prone to drought.

Turning to pest management, traditional chemical pesticides pose significant environmental and health risks. Innovations in biopesticides offer safer alternatives. These formulations, derived from natural materials like plant extracts and beneficial microorganisms, are less harmful to the ecosystem. Nanoencapsulation techniques enhance the efficacy of biopesticides, ensuring they are released in a targeted manner, reducing the need for frequent applications.

The integration of Internet of Things (IoT) with advanced materials is also transforming agriculture. Smart sensors made from innovative materials can monitor soil moisture, nutrient levels, and crop health in real-time. These sensors collect data that informs precision farming practices, allowing farmers to make more informed decisions about irrigation, fertilization, and pest control. This data-driven approach enhances resource efficiency and boosts crop yields while minimizing environmental impact.

One cannot overstate the potential of microbial innovations in promoting sustainable agriculture. Soil biota, including bacteria and fungi, play a critical role in nutrient cycling and soil health. Advances in microbial inoculants, which introduce beneficial microbes to the soil, can significantly improve crop resilience and productivity. These inoculants enhance the natural processes like nitrogen fixation and phosphorus solubilization, reducing the dependence on synthetic fertilizers.

We must also consider the role of genetically engineered materials. Innovations like CRISPR technology allow for precise genetic modifications in crops, enabling them to utilize resources more efficiently and withstand harsh environmental conditions. While this area is still under rigorous ethical and regulatory scrutiny, its potential to revolutionize agriculture cannot be ignored.

Beyond individual materials, the holistic integration of these innovations into farming systems is crucial. Sustainable agricultural practices are most effective when they work synergistically. For instance, combining biochar with nanofertilizers and IoT-based monitoring can create a closed-loop system where soil health, water efficiency, and crop productivity are all optimized simultaneously.

It's essential to foster collaboration between scientists, farmers, and policymakers to ensure the successful adoption of these innovative materials. Farmers need to be educated about the benefits and application methods, while policymakers must create conducive environments for research and implementation. Incentives and subsidies for sustainable practices can also accelerate this transition.

A key lesson from these innovations is that technology can provide solutions, but they must be implemented wisely and ethically. Over-reliance on any single material or technique can lead to new ecological imbalances. A diverse and integrated approach is the way forward, combining traditional wisdom with cutting-edge science.

Ultimately, promoting sustainable agriculture through innovative materials can pave the way for a food-secure future. These advancements are not mere supplements to existing practices but represent a paradigm shift toward a more resilient and sustainable approach to agriculture. When combined with a commitment to sustainability and ethical stewardship, innovative materials can indeed transform how we grow food and care for our planet.

By embracing these advances, we are not just improving farming practices; we are nurturing the future of humanity. Sustainable agriculture, powered by innovative materials, holds the promise of a healthier planet, richer soil, abundant crops, and a brighter future for generations to come. Let's set the groundwork today so that tomorrow's harvest is not only plentiful but also kind to the Earth.

Chapter 15:
The Human Element: Education and Ethical Considerations

In a world of ever-evolving technologies and pressing environmental challenges, the human element becomes a cornerstone for sustainable progress. Preparing the workforce of tomorrow goes beyond technical proficiency—it demands a robust foundation in ethical considerations that address the complex interplay between innovation and resource stewardship. We must educate individuals not only about the cutting-edge materials shaping our future but also about the values and principles guiding their use. As we navigate through the labyrinth of technological advancement, it's imperative to instill an ethical compass that prioritizes sustainability, equity, and social responsibility. Fostering a culture of continuous learning and ethical mindfulness will enable us to harness the power of technology while safeguarding our planet's finite resources for generations to come. The road ahead isn't just paved with new materials and technologies; it's built on the human capacity to make thoughtful, informed choices that align with a vision of a sustainable and ethical world.

Preparing the Workforce of Tomorrow

In an era where technological advancements are reshaping industries at a breakneck pace, preparing the workforce of tomorrow isn't just a priority—it's a necessity. Key to this preparation is a multifaceted educational framework that not only equips individuals with the technical skills required for new and emerging fields but also embeds a deep understanding of sustainability and ethical considerations. By fostering strong industry-academic partnerships, we can create dynamic curricula that respond to real-time challenges and opportunities, instilling a mindset that values innovation and environmental stewardship equally. Equally crucial is lifelong learning; providing continual training and development opportunities ensures that workers can adapt to evolving technologies and methodologies. At its core, this holistic approach to workforce development aims to create a generation of professionals who are not only proficient in cutting-edge technologies but are also conscious custodians of the planet's resources. By integrating forward-thinking education strategies and ethical foresight, we're not just preparing individuals for jobs—we're preparing them to lead the charge towards a sustainable and prosperous future.

The Ethics of Resource Usage and Technological Advancement sit at the very heart of our pursuit for a sustainable future. When we consider the intricate dance between technological development and resource consumption, we must grapple with complex ethical questions. How do we balance economic progress with environmental stewardship? What responsibility do we bear towards future generations in managing finite resources?

One of the most pressing ethical dilemmas is the disproportionate burden resource extraction places on certain communities. Often, the most vulnerable populations are located in regions rich in coveted materials such as cobalt, lithium, and rare earth elements. These communities face environmental degradation, public health concerns, and inequitable economic dynamics. Ethical resource usage mandates that corporations and governments engage in fair-trade practices, ensure safe working conditions, and reinvest in these communities to foster sustainable development.

The ethics of resource usage isn't only about extraction; it's also tied to the lifecycle of materials. We have a moral obligation to consider the environmental impact of materials throughout their entire lifecycle – from mining and production to use and disposal. This approach encourages strategies like recycling, material recovery, and the development of circular economies, which aim to maximize resource efficiency and minimize waste.

Technological advancements themselves must be scrutinized for their ethical implications. Take, for example, the electric vehicle (EV) revolution. While EVs represent a leap towards reducing fossil fuel dependency, their batteries rely heavily on lithium and cobalt, materials often extracted under ethically dubious conditions. Engineering innovations must therefore focus on creating alternative, less harmful materials and improving battery recycling methods.

The ethical use of technology also touches upon energy consumption. Advanced technologies, from data centers to block-chain, require enormous amounts of energy. Harnessing renewable energy sources such as solar and wind power to meet these needs is not just a technological challenge but an ethical imperative. Only by aligning technological growth with sustainable energy can we hope to mitigate the adverse impacts of our digital age.

Moreover, transparency and accountability are cornerstones of ethical resource management. Corporations must adopt transparent supply chains that allow consumers to make informed choices. Blockchain technology, for instance, could revolutionize how we trace the origin and journey of raw materials, ensuring every step aligns with ethical standards. Such transparency empowers consumers and holds companies accountable, fostering a culture of ethical consumption.

Intellectual property rights (IPR) also play a crucial role in the ethics of technological advancement. While patents and copyrights protect innovations, they can also stifle the dissemination of sustainable technologies. Striking a balance between protecting inventors' rights and ensuring broad access to green technologies is essential. Governments and international bodies must work to create frameworks that promote open sharing of critical advancements without undermining the incentive to innovate.

Democratization of technology is another ethical consideration. As new technologies emerge, there's a risk that they will be accessible only to affluent segments of society, further deepening economic and social divides. Ensuring that advancements in sustainability are accessible to all demographics is fundamental. This means prioritizing affordable and scalable solutions that can be adopted globally, particularly in developing nations.

Furthermore, education and awareness are vital for ethical resource usage. Tomorrow's leaders and consumers must be equipped with the knowledge and principles needed to make sustainable choices. Integrating sustainability and ethics into educational curricula at all levels can foster a generation that values and practices responsible resource management.

Corporate social responsibility (CSR) is yet another dimension where ethics intersect with resource usage and technological advancement. Companies today are expected to look beyond profits and address their environmental and social impacts. Genuine CSR initiatives go beyond greenwashing; they involve setting ambitious sustainability goals, reducing carbon footprints, and investing in community well-being.

Intergenerational equity also frames our ethical considerations. We need to think about the kind of planet we'll leave for future generations. Resource depletion and environmental degradation compromise the ability of future inhabitants to meet their own needs. As custodians of Earth, our ethical duty extends beyond immediate gains to ensuring a habitable and thriving world for posterity.

Ethical frameworks and standards, though sometimes perceived as constraints, actually drive innovation. When companies and researchers operate with an ethical compass, they're often inspired to discover novel, sustainable solutions. This kind of innovation doesn't just solve problems – it anticipates them, proactively seeking ways to circumvent ethical quandaries before they arise.

Global cooperation is indispensable to ethical resource usage. Resources are unevenly distributed across the planet, while the impacts of their misuse know no borders. Collective international action, shared knowledge, and cooperative policymaking are essential to ensuring that resource usage is fair, sustainable, and ethical.

Finally, ethical considerations must evolve with technological advancements. As new frontiers like artificial intelligence and biotechnology emerge, the ethical landscape will undoubtedly change. Staying ahead requires continuous evaluation and adaptation of our ethical standards to keep pace with innovation, ensuring that progress does not come at undue cost.

The Ethics of Resource Usage and Technological Advancement illustrate the interconnectedness of our decisions today with the sustainability of tomorrow. By embedding ethics into every facet of resource management and technological development, we move closer to a future where innovation and sustainability are not mutually exclusive, but mutually reinforcing.

Chapter 16:
Policy and Governance
for a Sustainable Future

Crafting effective policies and governance structures is essential for steering us towards a sustainable future. As global demands for critical resources swell, the urgency for innovative and adaptable policy frameworks becomes more evident. We must foster international cooperation and coordination, ensuring that nations work together rather than compete in ways that exhaust our planet's finite resources. Policies need to be forward-thinking, not just addressing the symptoms of resource scarcity but tackling its root causes. This means incentivizing sustainable practices, investing in technological advancements, and emphasizing the importance of ecological balance in every governmental decision. Governments and institutions should prioritize transparent, accountable, and inclusive policy-making processes that encourage public participation and stakeholder engagement. Embracing such comprehensive governance strategies will not only guide us in managing resources more efficiently but also inspire new generations to consider sustainability as a fundamental part of economic and technological progress. Together, we can envision a world where policy and innovation converge to create an enduring and prosperous future for all.

Shaping Policies for Sustainable Resource Management

To achieve a truly sustainable future, creating effective policies for resource management is crucial. By integrating environmental, economic, and social dimensions, policymakers can guide societies toward a balanced utilization of natural resources while minimizing ecological disruption. A broad, multidisciplinary approach is required, linking legislation, technological innovation, and societal behaviors. This holistically crafted policy framework ensures resources like rare earth elements, water, and energy are managed efficiently, limiting waste and promoting reuse. Policymakers must also commit to ongoing international collaboration, recognizing that resource scarcity and environmental sustainability are global challenges requiring unified and strategic responses. Impressive strides in recycling initiatives and green technologies could set the precedent, inspiring both developed and developing nations to adopt stringent yet flexible regulatory practices. The aim is not just to impose rules but to cultivate an ethos of responsible stewardship and innovation, ensuring these precious resources continue to sustain and advance future technologies.

International Cooperation on Resource Scarcity and Innovation is more than just a strategic necessity; it's an ethical imperative. At a time when the world faces unprecedented challenges in resource scarcity and environmental sustainability, nations must come together to address these issues collectively. The scarcity of resources such as rare earth elements, lithium, and clean water is not confined within borders. It's a global problem that requires a global response, powered by innovation and guided by a shared commitment to sustainability.

Resource scarcity, by its very nature, creates inequalities and exacerbates geopolitical tensions. Different countries possess varying levels of natural resources, technological capabilities, and economic power. These disparities can lead to conflict and competition rather than cooperation. Therefore, an international framework for resource management and innovation is essential. Think of it as countries forming a collective pact, not just to share resources, but to share knowledge, technology, and best practices.

Collaboration on resource scarcity and innovation can unfold in various ways. One key avenue for cooperation is in research and development (R&D). When countries pool their intellectual and financial resources, they can achieve breakthroughs more quickly and efficiently. Joint research initiatives could, for instance, lead to the development of new materials and technologies that are more sustainable and efficient. Such cooperation doesn't just accelerate innovation; it also ensures that the benefits of those innovations are distributed more equitably.

International organizations like the United Nations and the International Energy Agency have roles to play, too. They can act as conveners and facilitators, bringing together diverse stakeholders to hammer out agreements, set global standards, and monitor compliance. For instance, the Paris Agreement on climate change serves as a framework for global cooperation on reducing greenhouse gas emissions. Similar accords can be envisioned for the sustainable management of critical materials.

Another example is the European Union's Horizon 2020 program, which encourages multinational research collaborations to address societal challenges, including resource scarcity. Such programs can serve as templates for other regions to emulate and adapt to their contexts. By financing collaborative projects focused on sustainability, these initiatives can drive both innovation and cohesion across borders.

Interestingly, cooperation can also extend to sharing intellectual property (IP). Typically seen as a competitive edge, sharing patents and technological know-how in specific areas can accelerate the adoption of critical technologies worldwide. For instance, opening up patents related to renewable energy technologies can help developing nations leapfrog to cleaner energy solutions without reinventing the wheel.

However, international cooperation isn't just about sharing resources and technology. It's equally about developing shared norms and principles. Setting common environmental standards, for example, helps ensure that progress in one part of the world isn't undercut by setbacks elsewhere. A unified approach to regulations can make it easier for countries to comply and for enforcement to be more systematic and effective.

It's vital to involve a diverse range of stakeholders in these cooperative ventures, including governments, private sector companies, non-profit organizations, and academic institutions. Each of these actors brings unique capabilities and perspectives to the table. Governments can provide policy frameworks and funding; the private sector can drive innovation and scalability; non-profits can focus on advocacy and community engagement; and academic institutions can contribute cutting-edge research.

Balancing the interests of these diverse stakeholders requires skillful negotiation and diplomacy. But when done right, it can lead to synergistic relationships where the whole is greater than the sum of its parts. For example, the Global Battery Alliance, a multi-stakeholder initiative, aims to establish a sustainable battery value chain. Such collaborative models can be replicated to address other resource challenges.

One concrete step towards fostering international cooperation is through bilateral and multilateral agreements that explicitly focus on resource sustainability and innovation. These agreements can provide a formal structure for collaboration, specifying roles, responsibilities, and mechanisms for dispute resolution. They can also set clear, measurable targets and timelines, ensuring that all parties are held accountable.

Public awareness and education are also crucial components of international cooperation. Citizens around the world need to understand the importance of sustainable resource management and the steps that are being taken to achieve it. Media campaigns, educational programs, and grassroots movements can all play a role in raising awareness and fostering a culture of sustainability.

Funding is another critical aspect. International financial institutions like the World Bank and the International Monetary Fund can provide funding and technical assistance to countries that need help implementing sustainable practices. They can also facilitate public-private partnerships that leverage private sector investment for public good. Additionally, global investment funds focused on sustainability can channel capital towards projects that address resource scarcity and promote innovation.

It's worth noting that while cooperation is essential, it's not without its challenges. Differences in political ideologies, economic interests, and cultural values can make collaboration complex. Nonetheless, the urgency of addressing resource scarcity and promoting innovation makes overcoming these hurdles essential. Countries need to approach these challenges with flexibility, openness, and a willingness to compromise.

In summary, international cooperation on resource scarcity and innovation is a multifaceted endeavor that requires coordinated action across various levels. From shared research initiatives and collaborative frameworks to public education and financial support, there are numerous avenues through which nations can work together. By doing so, they can not only mitigate the challenges of resource scarcity but also unlock new opportunities for sustainable development and technological advancement. As we look to the future, such cooperation will be key in building a world that is both innovative and sustainable.

Chapter 17:
Investment and Economics
of New Materials

As we stand on the precipice of a technological renaissance, investment in new materials isn't just a choice, it's a necessity. The economics of innovation hinge on our ability to fund research and development into sustainable materials that can support the growing demand for clean energy, efficient manufacturing, and lessened environmental impact. Companies and governments must pivot toward strategies that not only reward short-term gains but also foster long-term sustainability. This shift is already underway, with venture capitalists and institutional investors recognizing the immense potential in sectors like battery technology, advanced composites, and recyclable polymers. The ripple effect of these investments can drive down costs, making these new materials more accessible and promoting widespread adoption. In essence, aligning financial incentives with environmental stewardship can create a win-win scenario where profitability meets sustainability. The economic landscape of the future will largely be determined by how well we marshal resources into these groundbreaking technologies today. Investing wisely now can lead to substantial payoffs, both in terms of financial returns and the health of our planet.

The Business of Sustainability

In an increasingly resource-conscious world, the business of sustainability transcends mere compliance; it represents a transformative approach to how we conceive, produce, and utilize materials. Embracing sustainability isn't just a moral imperative; it's an economic opportunity that reverberates across all industries. Companies that integrate sustainable practices at their core can see a significant return on investment through cost savings, innovation, and enhanced brand loyalty. Sustainable business models foster resilience, as they're designed to adapt to fluctuating resource availability and regulatory landscapes. Furthermore, investing in sustainable materials goes hand-in-hand with pioneering technologies, driving the market towards a greener, more efficient future. As stakeholders—from investors to consumers—gravitate towards eco-friendly solutions, the businesses that lead the charge in sustainable innovation will not only thrive but set the standard for future economic paradigms.

Funding the Future: Investment in Sustainable Technologies represents a key intersection where financial acumen meets environmental stewardship. Capturing the essence of sustainable progress, it requires not just recognizing the innovative technologies poised to shape our tomorrow, but also strategically funding these endeavors to ensure they come to fruition. Let's delve into how investments in sustainable technologies are laying the groundwork for an environmentally conscious and innovative society.

Investment in sustainable technologies isn't just a financial decision; it's a commitment to advancing society while safeguarding our planet. The transition toward renewable energy, eco-friendly materials, and sustainable practices can't happen without substantial financial backing. Investors, whether private entities, governments, or non-profits, have a pivotal role in nurturing early-stage innovations and scaling up proven solutions.

One of the most compelling reasons for investing in sustainable technologies is their potential for high returns. From breakthroughs in solar panel efficiency to advancements in wind turbine materials, these technologies promise not only environmental benefits but also economic value. Renewable energy projects, for instance, often provide stable, long-term returns which can appeal to risk-averse investors looking for steady income streams.

The financial landscape around sustainable technologies is rapidly evolving. Green bonds, sustainability-linked loans, and venture capital dedicated to clean tech are just a few examples of how the traditional financial instruments are being redefined to support environmental goals. These mechanisms are designed to mitigate financial risks associated with new technologies while promoting transparency and accountability in how funds are used.

Moreover, government policy and regulatory frameworks play a critical role in shaping investment patterns. Subsidies, tax incentives, and grants are powerful tools used by governments worldwide to encourage investments in renewable energy and sustainable materials. In many cases, these policies make the difference between a project being economically viable or not.

Aside from policy, public perception and consumer demand also drive investment toward sustainability. As consumers become more environmentally conscious, companies that adopt sustainable practices benefit from enhanced brand loyalty and a stronger market position. This shift in consumer preferences creates an added incentive for investors to fund green technologies that align with public sentiment.

Interestingly, the rise of Environmental, Social, and Governance (ESG) criteria in investment decisions signifies a paradigm shift. Investors are increasingly considering the broader impact of their investments, looking beyond mere financial returns to include social and environmental outcomes as critical performance indicators. ESG metrics thus facilitate a holistic approach to investment, marrying profit with purpose.

Sustainable technologies are not limited to renewable energy alone. Innovations in areas like biodegradable materials, water purification systems, and efficient waste management are equally important. Investing in these technologies addresses some of the most pressing environmental challenges, from plastic pollution to water scarcity, thereby contributing to comprehensive sustainability.

Venture capital plays a vital role in this ecosystem, often providing the crucial seed funding and mentorship required for startups to develop their concepts into market-ready products. These early-stage investments carry higher risks but offer the potential for significant impacts if the technologies succeed. The willingness of venture capitalists to take these risks is instrumental in bringing transformative technologies to light.

Moreover, collaborative efforts among stakeholders amplify the impact of investments. Private-public partnerships, industry consortia, and cross-border collaborations have proven effective in pooling resources, expertise, and networks to drive significant technological advances. Such collaborative funding models democratize access to sustainable technologies and ensure a broader distribution of their benefits.

The involvement of large corporations can't be overlooked. Multinational companies are increasingly investing in sustainable technologies both to future-proof their operations and to meet regulatory obligations and consumer expectations. Corporate venture arms and dedicated sustainability funds are becoming commonplace, signaling a strong commitment to integrating sustainability into core business strategies.

A crucial aspect of funding sustainable technologies is addressing the supply chain and circular economy models. Investments here enable the development of technologies that reduce waste, increase resource efficiency, and promote the reuse and recycling of materials. These technologies ensure that resources are utilized to their full potential, thereby reducing the environmental footprint and fostering economic resilience.

Educational institutions and research organizations also serve as critical conduits for innovation in sustainable technologies. Funding academic research and supporting university-industry partnerships stimulate cutting-edge discoveries and expedite their migration from labs to the marketplace. This bridge between academia and industry is essential for translating theoretical research into practical, scalable solutions.

Looking ahead, the democratization of investment in sustainable technologies through crowdfunding and community investment initiatives is beginning to reshape the landscape. These platforms enable everyday citizens to contribute financially to projects dedicated to sustainability, thereby spreading both the risks and rewards while fostering a shared sense of purpose and involvement.

Funding the Future: Investment in Sustainable Technologies is about more than just finance; it's about creating a vision for a sustainable world and mobilizing the necessary resources to make that vision a reality. By strategically targeting investments, aligning incentives with environmental goals, and fostering collaboration among diverse stakeholders, we can pave the way for technologies that not only promise a cleaner planet but also offer substantial economic opportunities.

Chapter 18:
Future Cities: Building Sustainable Urban Landscapes

As we envision the cities of tomorrow, the necessity for innovative solutions that marry technological advancement with ecological responsibility becomes paramount. The integration of smart materials in urban infrastructures isn't just a possibility—it's a necessity for fostering environments where efficiency, resilience, and sustainability coexist harmoniously. Future cities will harness data-driven urban planning to optimize everything from energy use to waste management, reducing our carbon footprint and making urban living more sustainable. Green spaces will play a pivotal role, acting as the lungs of these metropolitan areas, improving air quality, and providing havens for biodiversity amidst concrete jungles. By reimagining our urban landscapes with a focus on sustainability, we take crucial steps toward mitigating climate change, conserving resources, and ensuring a better quality of life for future generations. Every building, park, and street will need to reflect a commitment to a sustainable lifestyle, incorporating renewable energy sources, and sustainable materials, and fostering community engagement. The path ahead is challenging, but the rewards—cleaner air, better resource management, and a healthier planet—are invaluable.

Smart Materials for Smart Cities

In the race to build sustainable urban landscapes, smart materials play a pivotal role by transforming conventional city infrastructure into adaptive, efficient, and eco-friendly systems. These materials, often embedded with sensors and responsive capabilities, allow buildings, roads, and bridges to interact dynamically with their environments. Imagine sidewalks that generate electricity from footsteps or buildings that adjust their temperatures based on occupancy and external weather conditions. This level of innovation doesn't just reduce energy consumption; it fundamentally shifts how we experience and interact with urban spaces. By integrating intelligent materials into city planning, we can create urban ecosystems that are not only sustainable but also self-sustaining. This is the vision of future cities—where technology and materials converge to create living landscapes that respond to the needs of their inhabitants and the planet, heralding a new era of conscious urban development.

Urban Planning and the Role of Green Spaces seamlessly integrates with the broader narrative of sustainable urban development. As cities rapidly expand, the inclusion of green spaces becomes increasingly critical—not just as aesthetic luxuries, but as fundamental components of urban infrastructure. These areas offer a multifaceted solution to numerous environmental, social, and health issues, standing as pillars for a sustainable and technology-driven future.

Green spaces contribute significantly to ecological balance within urban environments. They help filter pollutants, improve air quality, and act as carbon sinks. By absorbing carbon dioxide and releasing oxygen through photosynthesis, urban parks and gardens play a vital role in mitigating climate change. This ecological service alone makes a compelling case for integrating green spaces into urban planning strategies.

But the role of green spaces extends far beyond improving air quality. They also play a crucial part in managing urban water cycles. Trees and plants help in absorbing rainwater, reducing surface runoff, and thus mitigating the risks of urban flooding. This natural water management system reduces the burden on city drainage systems and helps replenish groundwater resources, contributing to a more resilient urban habitat.

Equally important are the social benefits that green spaces bring to urban societies. They provide a venue for community interaction, outdoor activities, and recreation, enhancing the quality of life for residents. Research has shown that access to green spaces is linked to improved mental health, reduced stress, and overall well-being. These areas serve as communal grounds, fostering social cohesion and a sense of belonging among urban dwellers.

The inclusion of green spaces in urban planning is also immensely beneficial for biodiversity. They create habitats for various flora and fauna, contributing to urban biodiversity. Green corridors can help mitigate the fragmentation of landscapes, allowing wildlife to thrive even in bustling city environments. This not only enriches urban biodiversity but also promotes ecological education and awareness among city residents.

Urban heat islands—areas with significantly higher temperatures than their rural surroundings due to human activities—can be effectively countered with green spaces. Vegetation provides shade and releases moisture into the air, cooling the urban microclimate. This cooling effect not only improves comfort levels for residents but also reduces the energy demand for air conditioning, leading to substantial energy savings and lower greenhouse gas emissions.

Green spaces also enhance the aesthetic appeal of urban landscapes, which can boost property values and attract investment. An appealing environment encourages people to spend time outdoors, participating in physical activities that promote health and wellness. This, in turn, can reduce healthcare costs and foster a culture of active living.

Integrating green spaces into urban planning is not without challenges. It requires a thoughtful balance between development and conservation, necessitating cooperation between policymakers, urban planners, and citizens. Innovative approaches such as vertical gardens, green roofs, and urban agriculture are emerging as viable solutions to these challenges. These innovations not only maximize the use of limited urban space but also contribute to sustainable food production and local food security.

Moreover, policies that incentivize the creation and maintenance of green spaces are crucial. Urban municipalities can offer tax breaks or subsidies to developers who incorporate green spaces into their designs. Community engagement programs can also play a pivotal role by involving residents in the cultivation and preservation of these areas, ensuring long-term sustainability and a sense of collective ownership.

Educational initiatives can further amplify the impact of green spaces. Schools, colleges, and community centers can integrate environmental education into their curricula, highlighting the importance of green spaces in urban ecosystems. Citizen science projects and volunteer programs can also engage the community in the active monitoring and management of these areas, fostering a culture of environmental stewardship.

Technology offers exciting possibilities for enhancing the role of green spaces in urban planning. Advanced Geographic Information Systems (GIS) and remote sensing technologies can assist in optimal site selection and the efficient management of green spaces. Drones and IoT sensors can monitor plant health, soil conditions, and water needs, ensuring timely intervention and maintenance.

In envisioning the future of our cities, green spaces are not optional add-ons but essential components of sustainable urban ecosystems. They support environmental resilience, enhance social well-being, and promote economic vitality. As we march towards an era defined by technological advancements and sustainability, the integration of green spaces into urban planning will be instrumental in shaping livable, resilient, and thriving cities.

We must remember that the creation of green urban landscapes is a collaborative effort that requires visionary leadership, community involvement, and innovative technologies. By prioritizing green spaces in urban planning, we can cultivate environments that are not only sustainable but also enriching and harmonious, striking a balance between progress and preservation.

Urban planners, policymakers, and citizens must work together to ensure that green spaces are given the attention and resources they deserve. It is through this collaborative effort that we can build sustainable urban landscapes, fostering an environment in which technology and nature coexist harmoniously.

As cities continue to grow, the need for sustainable urban development strategies becomes even more pressing. Green spaces must be at the heart of these strategies, serving as a testament to our commitment to building a better, greener future for all. The role of green spaces in urban planning is thus a symbol of hope and a beacon for sustainable progress in the years to come.

Chapter 19:
Transportation: Driving
Towards Sustainability

As we steer into the future, the demand for sustainable transportation is at an all-time high, pressing us to innovate or face environmental calamities. While electric vehicles (EVs) have already begun revolutionizing our roads, the quest doesn't end there. We must explore alternative fuels for aviation and maritime transport to mitigate carbon footprints comprehensively. Hydrogen, biofuels, and advanced battery technologies are paving the way towards greener horizons. The marriage of cutting-edge materials and renewable energy sources promises a significant reduction in pollution and resource depletion. It's an era where sustainability isn't just a choice but a necessity, requiring collective efforts to transcend traditional boundaries, rethink mobility, and embrace a cleaner, smarter, and more efficient transportation ecosystem.

Electric Vehicles and Beyond

The electrification of transportation marks a monumental step in the quest for sustainability, transcending traditional combustion engines and dramatically reducing greenhouse gas emissions. Electric vehicles (EVs) are at the forefront of this revolution, leveraging advanced battery technologies to enhance their range, performance, and affordability. However, the impact of EVs extends far beyond personal transportation; they also influence public transit systems, freight logistics, and even air travel through nascent electric airplanes. As we strive to integrate these technologies into our everyday lives, the demand for critical materials like lithium, cobalt, and rare earth elements surges, prompting a more robust emphasis on ethical sourcing and recycling innovations. This paradigm shift necessitates a reimagined infrastructure, including widespread charging networks and smart grids, to support a seamless transition. In embracing electric mobility, we not only foster a greener planet but also pave the way for future advancements in transportation that align with our sustainability goals. The journey towards sustainable transportation is multidimensional, involving intricate balances between technological innovation, resource management, and environmental stewardship.

Sustainable Aviation and Marine Fuels are pivotal topics in our discussion of the future of transportation.

As we face the need to curb carbon emissions and protect our environment, the aviation and marine sectors stand at a crossroads. The journey toward sustainable fuels for these industries is not only a technological quest but a moral imperative. These sectors account for significant greenhouse gas emissions, and switching to sustainable fuels is a game-changer. These fuels must meet the energy intensity requirements while being cleaner for the planet. It's a delicate balancing act that the world cannot afford to get wrong.

Sustainable aviation fuels (SAFs) have progressed significantly over the past decade. Unlike conventional jet fuel derived from fossil fuels, SAFs are produced from renewable resources. These resources include biomass, recycled waste materials, and even captured carbon. Some promising examples include biofuels made from algae, municipal waste, and agricultural residues. The aviation industry's goal is to achieve carbon neutrality by 2050, and SAFs play a crucial role.

There have been some exciting breakthroughs. Notably, companies are exploring synthetic fuels produced by combining carbon dioxide captured from the atmosphere with hydrogen derived from water. This approach has the potential to create a carbon-neutral cycle, as the carbon released during the combustion of these fuels would match the amount captured during production. Reducing reliance on fossil fuels also helps mitigate the economic volatility associated with oil markets.

The efficacy of SAFs has been put to the test in commercial flights. Airlines worldwide have conducted numerous experimental flights using blends of traditional jet fuel and SAFs. Some flights have even run on 100% SAF. The feedback has been promising. These tests have demonstrated that SAFs can seamlessly integrate with existing aircraft engines and fuel infrastructure without compromising performance or safety.

The marine industry presents its own set of challenges and opportunities for sustainable fuel solutions. Shipping is the backbone of global trade, but it is also a significant emitter of sulfur oxides, nitrogen oxides, and carbon dioxide. The International Maritime Organization (IMO) has set ambitious targets for reducing greenhouse gas emissions, prompting a wave of research into alternative fuels and propulsion technologies.

One such alternative is liquefied natural gas (LNG). It's been adopted by several shipping lines as a transition fuel, offering lower emissions compared to heavy fuel oil. However, LNG is not without its challenges. Methane leakage during extraction and transportation can offset some of its environmental benefits. More sustainable options are necessary to meet long-term decarbonization goals.

Biodiesel and renewable diesel are attracting attention as potential marine fuels. These fuels are made from organic materials like vegetable oils and animal fats. Much like in aviation, the ability to produce these fuels from waste and non-food-based sources is key to their sustainability. Biodiesel can be used in existing engines with little to no modification, making it a practical near-term solution.

Even more futuristic is the concept of using hydrogen and battery-electric propulsion for ships. Hydrogen can be utilized in fuel cells to generate electricity, emitting only water vapor as a byproduct. While this technology is still in its infancy for large-scale maritime applications, it holds immense promise. Electric propulsion, powered by renewable energy sources, is another avenue worth exploring, particularly for short-sea shipping and inland waterway vessels.

The shift to sustainable aviation and marine fuels does not come without its challenges. One significant barrier is the production scale. Current infrastructure for SAF and alternative marine fuel production is minuscule compared to the demand. Scaling up production requires substantial investment and innovation in supply chain logistics. Moreover, the cost of these fuels often exceeds that of conventional fuels. Financial incentives, policy support, and international cooperation are crucial to bridge this gap.

Policy initiatives like the Carbon Offsetting and Reduction Scheme for International Aviation (CORSIA) and the IMO's measures reflect the regulatory push needed to drive change. These frameworks reward early adopters of sustainable fuels and penalize heavy carbon emitters, thus creating an economic justification for transitioning.

While we're still at the early stages, collaboration between governments, industries, and research institutions is yielding fruitful results. Partnerships are forming to develop and test new fuel types, create efficient production processes, and establish best practices for adopting these fuels on a broader scale. Public perception and consumer demand for sustainable travel options further boost the momentum.

We must understand that sustainable aviation and marine fuels are not a silver bullet but part of a broader strategy to decarbonize transportation. This comprehensive approach includes advancements in propulsion technology, aerodynamic design, operational efficiencies, and regulatory policies fostering innovation. Long-term success will be determined by our ability to synchronize these efforts and maintain focus on sustainable development goals.

Looking ahead, the roadmap to a cleaner future in aviation and marine sectors is laden with hurdles and high stakes, but the prizes are equally lofty. With each milestone we achieve in developing and deploying sustainable fuels, we move one step closer to a world where travel and trade no longer come at the expense of our planet. The responsibility lies with all of us, from policymakers crafting the regulations and scientists pioneering new technologies to industry leaders willing to take bold steps towards a sustainable future. Sustainable aviation and marine fuels are fundamental pillars of transport decarbonization, and their time is now.

Ultimately, our commitment to sustainable aviation and marine fuels isn't just about reducing emissions. It's about envisioning a world where economic growth aligns with ecological balance, where technology supports nature rather than exploits it. This vision can inspire us to develop innovative solutions that ensure a viable, vibrant future for generations to come.

Chapter 20:
The Role of Artificial Intelligence and Big Data

As we navigate the complexities of optimizing resource use for a sustainable future, artificial intelligence (AI) and big data emerge as indispensable allies. These technologies enable us to analyze vast datasets with unprecedented speed and precision, uncovering patterns and insights that can drive more efficient resource management. For instance, AI algorithms can predict energy usage patterns in smart grids, optimizing electricity distribution and reducing waste. Similarly, big data analytics can identify inefficiencies in water usage, guiding efforts to conserve this vital resource. Beyond optimization, AI and big data facilitate innovation in recycling and waste management, turning what was once considered trash into valuable inputs for new products. By harnessing these powerful tools, we not only advance our technological capabilities but also take significant leaps toward creating a more sustainable and resilient world. The integration of AI and big data is more than just an enhancement of existing processes; it's a transformative force poised to redefine how we interact with and manage the critical resources that will shape our future.

Optimizing Resource Use Through Technology

We stand on the cusp of a revolutionary shift in how we manage and optimize our resource use, driven by the exponential growth of artificial intelligence (AI) and big data analytics. Cutting-edge algorithms can analyze vast troves of data in real time, offering unprecedented insights into resource allocation, consumption patterns, and inefficiencies. By leveraging machine learning, businesses and governments can predict demand with greater accuracy, minimize waste, and streamline supply chains. For instance, AI can dynamically adjust energy grids, ensuring that renewable sources are utilized to their fullest potential, reducing reliance on fossil fuels. Furthermore, smart sensors and IoT devices gather data continuously, feeding recommendations to optimize water usage in agriculture and other sectors. The promise of AI and big data isn't just in cutting costs or boosting efficiency—they also offer a tangible path to a more sustainable, resource-conscious future. Making informed decisions today using the intelligence we gather from these technologies can lay the groundwork for sustainable prosperity for generations to come.

AI in Recycling and Waste Management has heralded a new era where technology intertwines with environmental stewardship. The growing rate of waste production worldwide exerts pressure on traditional recycling systems, making it imperative to seek more efficient solutions. AI, with its analytical prowess and machine learning capabilities, promises to revolutionize recycling and waste management, ensuring that resources are not only optimized but also conserved for future generations.

The core advantage of AI in the recycling industry lies in its ability to process vast amounts of data rapidly. Traditional methods of sorting and recycling involve manual labor, which is not only slow but also prone to errors. AI-powered systems, on the other hand, utilize advanced algorithms to scan, identify, and sort waste materials with unparalleled accuracy. These systems incorporate computer vision technologies that can distinguish between various types of materials, such as plastics, metals, and paper, enhancing the efficiency of recycling plants.

One of the most compelling applications of AI in recycling is robotic automation. Robots equipped with AI-driven sensors can handle the sorting process much more effectively than human workers. These robots use machine learning to continually improve their sorting accuracy, leading to higher purity rates for recycled materials. This degree of precision reduces contamination, ensuring that recycled goods are of higher quality and therefore more valuable in the market.

AI also plays a critical role in optimizing waste management logistics. By analyzing data regarding waste production patterns, AI algorithms can predict the optimal times for waste collection, reducing the frequency of collections and saving fuel. Additionally, AI-driven route optimization ensures that collection trucks follow the most efficient paths, further minimizing environmental impact.

Beyond sorting and logistics, AI can dramatically transform how recycling programs are managed at the community level. Predictive analytics tools, driven by AI, can identify which areas generate the most waste and what types of waste are most common. This information allows municipalities to tailor education and policy interventions to specific neighborhoods, thereby increasing recycling rates and lowering overall waste generation.

Moreover, AI is fueling innovations in material recovery. Enhanced material recovery facilities (MRFs) are being developed with AI systems that can identify and extract rare or valuable materials from waste streams. This capability is particularly important for recovering electronic waste, which contains critical metals and components that are expensive and environmentally damaging to mine anew.

Deployment of AI in recycling is not just about the technology itself but also about creating a sustainable business model. The efficiency gains achieved by AI can significantly reduce operational costs for waste management companies. These savings can then be reinvested into more advanced recycling technologies, creating a virtuous cycle of improvement and sustainability.

Let's consider the environmental impact. Reduced contamination in recycling processes means less waste ends up in landfills or incinerators. This translates to lower greenhouse gas emissions and reduced dependence on raw materials, aligning perfectly with global sustainability goals.

The application of AI extends to waste-to-energy processes as well. AI systems can optimize the operation of waste-to-energy plants, ensuring that the maximum energy yield is obtained from the least amount of waste. This not only provides an additional energy source but also reduces the environmental burden of waste disposal.

As we transition into an era where circular economies are becoming the norm, AI stands out as an indispensable tool. By integrating AI into every stage of the recycling and waste management process, from collection to processing to final disposal, we can significantly enhance resource efficiency. This shift not only preserves natural resources but also drives innovations that can lead to the development of new materials and technologies.

However, the path to widespread AI adoption in recycling and waste management is not without its challenges. Initial investment costs can be high, and there is a steep learning curve for waste management personnel to effectively utilize these new technologies. Partnerships between governments, private companies, and academic institutions will be crucial in overcoming these hurdles and ensuring that AI integrates seamlessly into recycling systems.

Public perception and acceptance of AI technologies also play a vital role. Educating the general public about the benefits of AI in waste management can foster greater participation in recycling programs. When people understand that every item they recycle is being processed more efficiently and effectively thanks to AI, it can drive behavioral changes that further increase recycling rates.

Legal and regulatory frameworks will need to evolve to accommodate the rapid advancements in AI technologies. Ensuring data privacy and security, particularly when AI systems collect and analyze large amounts of data from residential and commercial waste streams, will be essential in building public trust.

Looking ahead, the potential for AI in recycling and waste management is immense. As machine learning algorithms become more sophisticated, we can anticipate even greater efficiencies in how waste is processed and recycled. Future AI systems could incorporate Internet of Things (IoT) devices to create a fully connected and responsive waste management network that catches inefficiencies before they become problems.

In conclusion, the incorporation of AI into recycling and waste management is not a distant dream but an emerging reality that promises to align industry practices with sustainability goals. It is a symbiotic relationship where technology propels us towards a cleaner, more resource-efficient future. As we embrace these advancements, we move a step closer to a world where waste is not a problem to be solved but a resource to be harnessed.

Chapter 21:
The Interconnectivity of
Global Resources

It's no longer feasible to examine resource management in isolation; the interconnectivity of global resources has become a defining characteristic of our era. Every decision made in one corner of the world can ripple across continents, influencing economies, ecosystems, and societies in profound ways. Imagine how a single technological advancement in recycling rare earth elements can alleviate geopolitical tensions, reduce environmental degradation, and foster economic stability in resource-scarce regions. To ensure sustainable access to these vital materials, collaborative international strategies and cross-disciplinary approaches are paramount. By acknowledging and acting on our interconnected reality, we unlock opportunities to innovate sustainably, ensuring that the benefits of natural resources extend to all corners of the globe. The task is monumental, but the convergence of technology, policy, and global cooperation offers a roadmap to a sustainable and equitable resource future.

The Global Impact of Local Decisions

Local decisions about resource management reverberate across the entire globe, illustrating the profound interconnectivity of our world. When a community opts for renewable energy sources over fossil fuels, it doesn't just reduce local pollution; it contributes to the global fight against climate change, setting an example that other regions may follow. Similarly, the extraction methods chosen by a single mining operation can influence international supply chains and economic stability, either perpetuating environmental degradation or encouraging sustainable practices. Every choice made on a micro-scale—from urban recycling programs to agricultural water use—has ripple effects that can bolster or undermine global sustainability efforts. It's a mosaic where each piece contributes to the larger picture, echoing through economies, ecosystems, and even geopolitical landscapes. Understanding this interconnectedness empowers individuals and policymakers alike to make more informed, strategic decisions that foster a sustainable future, realizing that their choices hold the potential to inspire substantial change far beyond their immediate surroundings.

Ensuring Access to Resources for All is paramount when considering the future of technology and sustainability. As we probe deeper into the 21st century, the equitable distribution of vital resources emerges not merely as a technical challenge but as an ethical imperative. With the rapid pace at which technological advancements occur, ensuring everyone has access to the resources that power these innovations is critical to fostering a just and inclusive society.

From advanced batteries to critical nanomaterials, the resources that will define the future are often limited and unequally distributed. This inequity can lead to resource monopolies, supply chain vulnerabilities, and geopolitical tensions. But more importantly, it can stall societal progress, affecting those who already face significant barriers to technological access. By championing fair resource distribution, we not only empower underrepresented communities but also pave the way for more resilient and innovative global systems.

The first step in this direction involves rethinking how we source and utilize these resources. Traditional mining and extraction methods, often plagued by inefficiency and environmental degradation, are not sustainable long-term solutions. Innovations in green mining technologies can significantly mitigate the environmental impact while optimizing the yield of critical materials. But greening the mining sector alone isn't enough.

Urban mining, the process of reclaiming raw materials from spent products, buildings, and waste, represents a burgeoning frontier. Cities are treasure troves of discarded electronics, offering a reservoir of high-value components waiting to be extracted and reused. By prioritizing urban mining, we can alleviate some dependence on traditional mining methods, democratize access to resources, and create a circular economy that sustains itself.

However, equitable resource distribution is not only about technology; it is also about governance and policy-making. Policy-makers across the globe must cooperate to ensure that resource-rich nations are not exploited and that benefits are shared equitably. International agreements can play a pivotal role in establishing transparent and fair trading practices, alongside commitments to sustainable management and environmental protection.

Optimizing resource use through technology is another critical factor. AI and big data can significantly enhance the efficiency of resource allocation, minimizing waste and maximizing utility. When properly harnessed, these technologies can provide invaluable insights into consumption patterns and predictive modeling to ensure resources are distributed where they are most needed. This ensures that even the remotest corners of the world can benefit from technological advancements.

Achieving this vision requires a collective effort, involving not just governments and corporations but also individuals. Education plays an essential part. We need to prepare a future workforce that understands the nuances of sustainable resource management and is equipped to drive innovation in this field. Educational institutions must incorporate sustainability modules across their curricula, ensuring that tomorrow's leaders are well-versed in ethical and sustainable practices.

Moreover, there's a pressing need to address the ethics of resource usage. It's crucial to strike a balance between fostering technological progress and ensuring that such progress doesn't come at the expense of marginalized communities or the environment. Ethical considerations, often sidelined in the race for technological superiority, should be integral to any discussion on future resources.

Communities can play a significant role in this transformation. Engaging local populations in resource management can democratize access and empower people to take charge of their futures. Community-led initiatives in recycling, waste management, and energy conservation can serve as beacons of sustainable living, showcasing the tangible benefits of equitable resource distribution.

Furthermore, investment in sustainable technologies needs to be prioritized. Venture capitalists and financial institutions should recognize the long-term value of investing in green technologies and sustainable resource management. Public and private sectors must collaborate to fund projects that aim at optimizing and democratizing resource usage, ensuring that innovations are accessible to all segments of society.

International cooperation remains a linchpin for success. Resource scarcity and environmental challenges are global issues that transcend borders. Collaborative frameworks, such as international research initiatives and cross-border partnerships, can foster innovation and ensure that advancements are widely shared and implemented.

One cannot overstate the importance of holistic urban planning in ensuring access to resources. Future cities must be designed with sustainability at their core, incorporating smart materials, green spaces, and efficient resource management systems. Urban designs should prioritize the well-being of all residents, reducing inequalities and ensuring that everyone benefits from technological advancements.

Lastly, the role of citizens in driving sustainable change cannot be overlooked. Individual actions, when scaled, can have a monumental impact. Whether it's through reducing personal consumption, supporting sustainable products, or participating in local clean-up drives, every small effort contributes to a larger movement towards equitable resource distribution.

In conclusion, **Ensuring Access to Resources for All** is an overarching goal that bridges numerous sectors and requires a multi-faceted approach. By leveraging technology, fostering education, promoting ethical practices, encouraging investments, and facilitating international cooperation, we can build a world where technology serves as a pathway to equality and sustainability rather than a divider.

Chapter 22:
Challenges on the Path
to Sustainability

Embarking on a sustainable future is fraught with numerous hurdles, as the intricate balance between technological advancements and environmental stewardship presents both opportunities and obstacles. Key challenges lie in overcoming technical barriers, such as developing materials that are both high-performing and environmentally friendly. Economically, the transition requires significant investment and a shift in market dynamics, which can be slow and resistant. Societally, there is the pressing need for widespread behavioral change and acceptance of new norms that prioritize sustainability over convenience. Integrating these elements demands innovative thinking and dedicated policy frameworks that support not just the creation but also the adoption of sustainable technologies. Despite these obstacles, the collective will to effectuate change—driven by heightened awareness and a commitment to future generations—fuels an unwavering optimism that we can indeed navigate through these challenges toward a more sustainable world.

Overcoming Technical and Economic Barriers

Navigating through the labyrinth of technical and economic barriers is essential for propelling sustainable technologies to the forefront. Technical obstacles like the complex processing of critical materials and the integration of renewable energy systems must be addressed with relentless innovation. On the economic side, the high initial costs often deter investment despite the long-term benefits. We must push for policy incentives, public-private partnerships, and innovative financing models to bridge the gap. In essence, a concerted effort combining cutting-edge research, strategic investments, and supportive policies can dismantle these barriers, paving the way for a future where technological advancements harmonize with environmental stewardship. This synthesis of technical prowess and economic viability is not just desirable—it's imperative for a sustainable tomorrow.

Societal Adaptations to a New Resource Paradigm mark a transformative moment in human history. We are on the cusp of a significant shift in how we source, manage, and utilize the raw materials that power our world. This transition isn't merely technical; it requires profound societal changes that will touch every aspect of our lives—from the economy to culture, from individual habits to systemic governance. Understanding and navigating this change is pivotal as we strive for a sustainable future.

The first step in societal adaptation is awareness. People need to comprehend the finite nature of traditional resources and the necessity to shift towards more sustainable alternatives. This involves education and public outreach programs, which will play a crucial role in altering societal mindsets. Educational institutions should integrate sustainability topics into their curriculums, emphasizing the importance of sustainable resource management from a young age. This foundation will enable future generations to make informed decisions, fostering a culture of sustainability.

Public policies must evolve to support this new resource paradigm. Governments will need to create regulatory frameworks that incentivize sustainable practices and penalize unsustainable ones. Policy shifts can drive corporations and individuals to innovate and adopt sustainable methods. For instance, tax incentives for the use of recycled materials and penalties for high carbon emissions could be potent tools. International cooperation on resource management will be crucial, as resource scarcity and environmental impacts do not recognize borders.

The business sector will also need to adapt, with companies rethinking their supply chains to prioritize sustainability. Businesses must embrace circular economy principles, designing products with their entire lifecycle in mind. This means creating goods that are durable, repairable, and recyclable. Transparency in sourcing and sustainability reporting will become essential, as consumers increasingly demand that companies align with their values. This trend will drive innovation and lead to the development of new markets focused on sustainable products.

In parallel, technological innovations will facilitate the transition to a new resource paradigm. Advances in renewable energy, recycling technologies, and sustainable materials will reduce our dependency on finite resources. However, technological solutions alone are not sufficient. Their adoption and efficient implementation require societal willingness and behavioral changes at both individual and collective levels. Investments in research and development are needed to ensure that these technologies are accessible and affordable.

Community initiatives will also play a vital role. Local governments and community groups can spearhead efforts such as urban farming, local recycling programs, and renewable energy projects. These initiatives not only reduce the community's environmental footprint but also foster a sense of collective responsibility and empowerment. Success stories from such community-led projects can inspire broader societal changes.

Transportation is one sector where societal adaptation will be particularly visible. As we shift towards electric vehicles and sustainable fuels, our daily commutes and travel habits will change. Infrastructure development, like the installation of charging stations, and policy incentives, such as subsidies for electric vehicles, will encourage people to make the switch. Public transportation systems will need to become more efficient and sustainable, offering convenient alternatives to single-passenger car travel.

Housing and urban planning must also adapt to this new paradigm. Future cities will need to incorporate smart materials and green spaces to optimize resource use and minimize environmental impacts. Urban planning policies should encourage the construction of energy-efficient buildings and the implementation of sustainable waste management systems. Residents will need to adapt to these changes, which might include adopting new habits like composting, energy-saving practices, and participating in community gardening.

On an individual level, lifestyle changes will be essential. People will need to adopt more sustainable consumption patterns, prioritizing products that are ethically sourced and environmentally friendly. This shift might involve reducing waste, recycling more effectively, and embracing a minimalist lifestyle that focuses on quality over quantity. Awareness campaigns and community programs can provide the necessary guidance and support for these lifestyle changes.

Equity and accessibility must be central to this transition. Ensuring that all segments of society have access to sustainable technologies and practices is crucial. This means that policies should aim to make sustainable options affordable and accessible to everyone, regardless of socioeconomic status. Efforts to reduce energy poverty, for instance, can involve subsidizing renewable energy installations for low-income households or providing financial assistance programs for energy-efficient home upgrades.

The workplace will also undergo significant changes. Industries reliant on traditional resources will need to pivot towards sustainable alternatives, potentially transforming job markets. Workforce development programs will be crucial to reskill workers and prepare them for new opportunities in green sectors. Emphasis on vocational training and continuous learning will ensure that the workforce can keep up with technological advancements and evolving industry demands.

Collaboration between governments, businesses, and civil society will be essential for a successful transition. Public-private partnerships can drive large-scale sustainability projects, from renewable energy installations to urban redevelopment plans. These collaborations can pool resources, expertise, and innovation to tackle the complex challenges of resource management.

Given the global nature of resource challenges, international cooperation will be vital. Countries need to work together to share best practices, develop global standards, and facilitate the transfer of sustainable technologies. International agreements and alliances can ensure that resource management strategies are cohesive and effective on a global scale, addressing both local and global environmental impacts.

In the face of these changes, resilience and adaptability will be key. Societies must be prepared to respond to unforeseen challenges and opportunities that arise during the transition. Flexibility in policies, openness to new ideas, and a willingness to experiment with innovative solutions will help society navigate the uncertainties of this paradigm shift.

Ultimately, the success of societal adaptations to a new resource paradigm hinges on our collective will and action. It requires a holistic approach that encompasses education, policy, business innovation, community involvement, lifestyle changes, equity, collaboration, and resilience. By embracing this holistic approach, we can create a sustainable future where resource management aligns with environmental stewardship and technological advancement, ensuring that the vital resources of the future are available for generations to come.

This transformation is not just about surviving in a resource-constrained world; it's about thriving sustainably and equitably. It's about turning challenges into opportunities and fostering a society that values and protects the planet's finite resources. It's about building a future where technological innovation and environmental sustainability are not mutually exclusive but are integral to our way of life.

Chapter 23:
Case Studies in Innovation
and Sustainability

In an increasingly resource-constrained world, real-world examples of how innovation and sustainability intersect serve as both inspiration and a roadmap for future efforts. Across continents, pioneering companies and communities illustrate the transformative power of sustainable practices, from Iceland's comprehensive utilization of geothermal energy to Kenya's extensive adoption of mobile banking to promote economic inclusiveness. These case studies reveal not only technological advances but also an underlying shift in mindset—valuing long-term impact over short-term gain. Examining these initiatives shows that with creativity and determination, formidable obstacles can be turned into opportunities for sustainable growth. The lessons learned from these success stories underscore the importance of collaboration, innovation, and adaptability in achieving a sustainable future for all. These examples beacon the potential, emphasizing that profound change is within reach when innovation is paired with an unwavering commitment to sustainability.

Success Stories from Around the World

Across every continent, nations and communities have made remarkable strides in the realms of innovation and sustainability, often against staggering odds. In Denmark, groundbreaking wind turbine technology has propelled the country into a global leader in renewable energy, setting a powerful example of how to harness nature's forces sustainably. Similarly, Kenya's adoption of mobile payment systems has significantly enhanced financial inclusion and supported local ecosystems by reducing paper currency dependency. Meanwhile, Sweden's comprehensive recycling program has dramatically reduced landfill waste, transforming trash into a valuable resource through innovative waste-to-energy processes. These stories illustrate the diverse potential for positive change, depicting a world where sustainable practices are not just possible, but already thriving. They offer invaluable lessons and serve as a testament to human ingenuity, showing us that with determination and innovation, a sustainable future is not merely a distant dream but an achievable reality.

Lessons Learned and Paths Forward in the context of resource sustainability are instrumental for guiding future innovation and policy. Our journey through the landscape of critical materials, renewable energy, and sustainable practices has yielded significant insights which are crucial for driving change. Understanding the multifaceted challenges and opportunities in resource management provides us with a roadmap to navigate the complexities ahead.

One major lesson is the importance of integrating sustainability into the very core of technological development. Throughout this book, we've explored how rare earth elements, lithium, and other materials are pivotal to modern innovations like electric vehicles and renewable energy systems. However, without sustainable sourcing and recycling methods, the very technologies that promise a greener future could contribute to environmental degradation. This paradox underscores the need for a comprehensive approach that balances technological progress with ecological responsibility.

Another profound takeaway is the significance of collaboration across sectors and borders. The global nature of resource distribution and its management demands international cooperation. Policies encouraging shared research, equitable resource distribution, and joint ventures in sustainable practices are essential. For instance, the interdependence in the supply of rare earth elements and the complexities in recycling lithium batteries highlight that no single nation can tackle these challenges in isolation.

Moreover, the advancements in renewable energy technologies, such as solar and wind power, have shown us the potential of innovation paired with sustainability. Yet, these advancements are not without their material challenges. The development of more efficient and resilient materials for solar panels, wind turbine blades, and energy storage solutions is key to maximizing the benefits of renewable resources. Innovative approaches, such as the use of perovskites for solar cells and composites in wind energy, pave the way for ongoing improvement.

Importantly, we've learned that the circular economy model offers compelling solutions to resource inefficiency and waste. By designing products with durability, reusability, and recyclability in mind, we can significantly reduce our environmental footprint. The shift towards a circular economy not only mitigates waste but also creates economic opportunities through new business models and job creation in recycling and remanufacturing industries.

In the realm of policy and governance, effective regulation and supportive frameworks are paramount. Policymakers play a crucial role in fostering an environment that encourages sustainable practices. This involves setting stringent standards for resource extraction, incentivizing green technologies, and ensuring that economic activities align with environmental goals. The synergy between government policies and market-driven innovations can drive meaningful progress.

Education and ethical considerations cannot be overlooked. Preparing the workforce for future technologies requires an emphasis on interdisciplinary knowledge, combining elements of engineering, environmental science, and social responsibility. Additionally, ethical considerations in resource usage and technological advancements need to be at the forefront to address issues like labor rights, environmental justice, and fair resource distribution.

Investments in sustainable technologies and practices are another cornerstone of our path forward. Funding is essential for research and development, scaling up new solutions, and overcoming initial economic barriers. Investments that prioritize long-term sustainability over short-term gains can lead to more resilient economies and healthier ecosystems. Encouraging venture capital, governmental grants, and public-private partnerships can be instrumental in driving forward sustainable innovations.

One of the most promising paths forward is urban planning and the role of green spaces in future cities. Smart cities, harnessing intelligent materials and technology for efficient resource use, can serve as beacons of sustainability. Urban areas, designed with integrated green spaces and eco-friendly infrastructure, not only enhance the quality of life but also contribute to climate resilience and biodiversity preservation.

Transportation systems are undergoing a transformative shift towards sustainability, with electric vehicles, sustainable aviation fuels, and innovations in shipping showing the way. However, the transition must go hand-in-hand with advancements in material science to ensure the sustainability of the resources involved. Research into alternative materials for batteries, lightweight composites for vehicles, and green fuels is critical to reducing the sector's carbon footprint.

Artificial intelligence (AI) and big data have shown immense potential in optimizing resource use and streamlining recycling processes. Leveraging AI can lead to more dynamic and efficient systems for managing resources, reducing waste, and enhancing supply chain transparency. Implementing these technologies could greatly increase our ability to meet sustainability targets.

On an individual level, promoting sustainable practices involves more than just technological or infrastructural changes. It requires a cultural shift toward valuing sustainability in everyday actions. This involves education on the environmental impacts of personal choices, encouraging participation in local sustainability initiatives, and fostering a mindset that prioritizes long-term ecological health over immediate convenience or cost savings.

As we contemplate future steps, the interconnectedness of global resources becomes increasingly evident. Local decisions often have wide-reaching impacts. Thus, fostering global cooperation and understanding is critical for equitable resource management. Ensuring access to vital resources for all communities, particularly those most vulnerable, is not just a technological or economic challenge but a moral imperative.

The barriers to sustainable resource management are numerous, whether technical, economic, or societal. Yet, they are not insurmountable. Through collective will, innovation, and commitment to sustainable principles, we can overcome these hurdles. Recognizing the sector-specific challenges and addressing them with targeted solutions will be key in achieving sustainable progress.

Finally, learning from past successes and setbacks will inform our future actions. Case studies from around the world provide valuable lessons on what strategies work and what missteps to avoid. By analyzing these examples, we can refine our approaches, adopt best practices, and avoid repeating past mistakes.

The path forward is clear: it's one of innovation, collaboration, and unwavering commitment to sustainability. Armed with the lessons learned, we can forge a future where technology and sustainability are not at odds but harmoniously aligned. This vision can inspire actions today that will shape a bright and sustainable tomorrow.

Chapter 24:
Visions of a Sustainable World

As we stand at the precipice of technological and environmental evolution, the vision of a sustainable world emerges as both a beacon and a blueprint. Envisioning a future where the synergy between resource management and innovative technologies flourishes is not just an exercise in imagination, but a plan rooted in actionable insights. In this future, advanced technologies are seamlessly integrated with natural processes, creating an ecosystem that nurtures while it innovates. The materials we rely on will be sourced and recycled responsibly, with minimal impact on our planet. Green energy will dominate, with solar and wind power leading the way, supported by efficient storage solutions. Water, air, and soil will no longer be spoiled treasures but safeguarded and revitalized, promoting a balanced interconnectivity of global resources. Above all, this vision underscores the imperative of cooperation—among nations, industries, and communities—to achieve a harmonious coexistence with our environment. It invites us to imagine not just what is possible, but what is necessary, inspiring us to move from vision to reality with urgency and purpose.

Imagining the Future of Technology and Resources

As we peer into the future, it's clear that the interplay between technology and resources will define how we live and thrive. The seamless integration of cutting-edge innovation with sustainable resource management isn't just aspirational; it's imperative for our collective future. From the marvels of smart materials reimagining the very fabric of our cities to energy matrices optimizing solar, wind, and perhaps even fusion, our path forward lies in the synergy of environmental stewardship and technological prowess. Imagining breakthroughs like efficient battery storage or quantum computing influencing resource logistics inspires not just scientific curiosity but actionable urgency. We are poised on the cusp of an era where collaboration, both international and interdisciplinary, drives holistic solutions. The work ahead is steeped in challenges, yet unequivocally laden with promise, inviting us to transform visionary concepts into tangible realities. This harmonious future isn't merely a distant dream—it's a beacon guiding us towards an ecologically balanced and innovatively abundant tomorrow.

From Vision to Reality: Steps We Can Take Today bridges the gap between our aspirations and actionable measures we can implement now. To turn our dreams of sustainable technology and resource management into tangible outcomes, we need to embrace a multifaceted approach. This involves a mixture of individual, corporate, and governmental actions, each playing a pivotal role. Let us delve into the practical steps we can take today to ensure the future we envision.

First and foremost, fostering a culture of innovation and sustainability in educational institutions is essential. Integrating sustainability and environmental science into curriculums at various educational levels helps cultivate a generation of environmentally conscious individuals. By prioritizing research in sustainable technologies and providing grants for projects that focus on resource efficiency, educational institutions can become hotspots for groundbreaking innovations that translate vision into reality.

Corporations must also re-evaluate their role in the sustainability landscape. Businesses should embrace the principles of a circular economy, where products are designed for durability, reuse, and recycling. Companies can start small by auditing their existing supply chains and identifying areas where sustainable alternatives can be integrated. An example could be using biodegradable packaging materials or shifting to energy-efficient production processes. These changes not only benefit the environment but also often result in cost savings in the long run.

On an individual level, consumers carry significant weight in driving sustainable practices through their purchasing power. By making informed choices, such as opting for products with minimal environmental impact and supporting companies that prioritize sustainability, consumers indirectly compel businesses to adopt greener practices. Simple changes in daily habits, like reducing plastic use, conserving water, and recycling, collectively make a substantial impact.

Governmental policies play a critical role in legislating sustainability. Governments can introduce and enforce regulations that mandate sustainable practices across various sectors. Policies such as tax incentives for green initiatives, stricter emission controls, and subsidies for renewable energy projects create a conducive environment for sustainable progress. It's vital for governments to work globally, cooperating with international bodies to address resource scarcity and promote innovation in resource management.

Investing in research and development is another crucial step. Funding innovative projects focused on renewable energy, advanced materials, and resource efficiency can lead to significant breakthroughs. Governments, private investors, and academic institutions should collaborate in creating research hubs that focus on the development and commercialization of sustainable technologies.

Community engagement should not be overlooked. Grassroots movements advocating for sustainability can significantly impact local practices. Communities can start initiatives such as local recycling programs, community gardens, and workshops on sustainable living. Engaging people at the community level not only fosters a sense of responsibility but also encourages collective action.

Adoption of renewable energy solutions needs to be accelerated. Solar panels, wind turbines, and other renewable energy sources should be prioritized over fossil fuels. Governments, businesses, and individuals can invest in renewable energy projects. Policies that lower the cost of renewable energy technologies and provide incentives for their adoption can expedite this transition.

Transport is another sector ripe for transformation. Promoting electric vehicles (EVs) and the requisite infrastructure, like charging stations, can reduce carbon emissions substantially. Governments can offer incentives for purchasing EVs and invest in public transportation systems powered by renewable energy. Encouraging the development and use of sustainable aviation and marine fuels also plays a part in this landscape.

Water conservation needs immediate focus. Innovations in water purification technologies and infrastructure investment to reduce water wastage are vital steps. Educating the public on conservation techniques and implementing water-saving measures in agriculture and industry can go a long way in ensuring water sustainability.

The agricultural sector must also evolve. Promoting sustainable agriculture practices through the use of innovative materials and soil remediation technologies helps protect one of our most vital resources—soil. Providing farmers with the knowledge and tools to implement these practices is a step towards long-term sustainability.

Encouraging urban mining, where electronic and other waste materials are recycled for rare and valuable materials, can reduce the strain on natural resources. Developing efficient recycling technologies and creating regulations that mandate recycling of electronic waste help close the loop in material usage.

Investing in clean nuclear energy, especially in thorium reactors, and managing nuclear waste effectively are other actionable steps. Bridging the gap between current nuclear technologies and future sustainable practices requires both policy support and technological innovation.

Public awareness campaigns about the importance of sustainability can change behaviors on a broader scale. Educational programs, advertising, and social media campaigns can drive home the importance of sustainable living and the steps individuals can take.

Finally, international cooperation is crucial. Global challenges require global solutions, and countries must work together to share technologies, knowledge, and resources. Collaboratively addressing resource management can ensure everyone benefits from the advancements in sustainability.

In summary, making the leap from vision to reality involves a diverse array of actions. Whether it's through education, corporate responsibility, individual actions, policy changes, or international cooperation, each step taken today paves the way for a sustainable future. The time to act is now, ensuring that our vision isn't just a dream but a vibrant, sustainable reality.

Chapter 25:
The Citizen's Guide to Contributing to a Sustainable Future

A s we've traversed the multifaceted landscape of technology and sustainability, it's clear that our individual contributions are pivotal to forging a sustainable future. From simple habits like reducing waste and conserving energy to more involved actions such as participating in local environmental initiatives, each step holds profound significance. Imagine the collective impact when communities globally prioritize eco-friendly practices; the ripple effect can lead to substantial shifts in demand for cleaner technologies and sustainable products. By educating ourselves and others, supporting policies that encourage resource efficiency, and opting for green alternatives, we embody the change we wish to see. The future belongs to those committed to innovation and environmental stewardship, proving that even small, consistent efforts matter. It's about a shared responsibility where every action we take nudges us closer to a sustainable and resilient world.

Everyday Actions for Sustainable Living

Every small action we take in our everyday lives can ripple out to create significant environmental impacts. By being mindful about our resource consumption and waste generation, we can contribute to a sustainable future in practical ways. For instance, opting for reusable bags, bottles, and containers minimizes plastic waste, while using energy-efficient appliances and lighting reduces your carbon footprint. Don't overlook the power of what you eat; incorporating more plant-based meals not only benefits your health but also conserves water and reduces greenhouse gas emissions. Additionally, supporting local and sustainable products helps create a demand for environmentally conscious goods and services, driving broader systemic changes. Lastly, consider advocating and voting for policies that promote renewable energy and resource conservation. These everyday actions—although simple—can collectively foster an environmentally conscious society, pushing us closer to sustainable technological advancements.

Engaging in Community and Global Efforts is essential for driving meaningful change towards a sustainable future. By fostering collaboration locally and globally, individuals and communities can contribute significantly to the management and optimization of resources. This engagement doesn't just connect us; it amplifies our impact.

Community efforts begin at home and extend outward. Simple acts such as reducing waste, conserving energy, and educating others about sustainable practices lay a foundation for broader influence. When single households adopt these habits, the ripple effect can spread across neighborhoods, cities, and eventually, nations.

Grassroots movements have historically been powerful catalysts for change. Environmental campaigns aimed at reducing plastic use or promoting renewable energy adoption are prime examples. These movements rely on the collective power of communities; when people unite for a cause, their combined voice can push policymakers and corporations towards sustainable practices.

Community gardens, for instance, not only provide local produce but also teach participants about sustainable agriculture and composting. This hands-on education can shift perceptions and create lifelong habits that contribute to resource efficiency and sustainability.

Schools and local governments can support such initiatives by providing space and resources.

Moreover, local businesses play a significant role in community sustainability. By sourcing materials responsibly and implementing eco-friendly practices, they set a standard for others. Businesses also serve as role models, illustrating the economic and environmental benefits of sustainability. Supporting these businesses reinforces a community's commitment to green practices.

Transitioning to global efforts, international cooperation is paramount. Climate change and resource scarcity are borderless issues; they require a unified global response. Organizations like the United Nations and the World Bank are pivotal in mobilizing resources and creating frameworks for sustainable development. Collaboration between countries can lead to innovative solutions and shared technologies.

Global efforts also include large-scale international agreements and treaties aimed at reducing carbon emissions and protecting bio-diversity. The Paris Agreement is a landmark example of an orchestrated effort to combat climate change. Such agreements signal a global acknowledgment of the need for collective action and provide a platform for accountability.

NGOs play a vital role in bridging the gap between government policies and community action. Organizations like Greenpeace and World Wildlife Fund raise awareness, mobilize volunteers, and advocate for policy changes. Their campaigns galvanize public support and hold governments and corporations accountable for their environmental impact.

Another critical avenue is international funding for sustainable projects. Financial institutions and development banks can provide loans and grants to nations and projects dedicated to sustainability.

Investing in renewable energy infrastructure or conservation projects in developing countries can make a significant difference globally.

Education and transparency are vital. Sharing knowledge and research across borders can expedite the adoption of best practices. Collaborative research initiatives can lead to breakthroughs in sustainable technologies and materials by combining intellectual resources from diverse backgrounds.

On a more individual level, global engagement involves conscious consumerism. Understanding the origins of the products we buy and choosing options that support fair trade and sustainable practices empower individuals as participants in the global movement towards sustainability. Every purchase becomes a part of a larger story of change.

Participation in global forums and social media campaigns can also amplify individual and community efforts. Platforms like these provide an opportunity to share successes, ask for support, and inspire others worldwide. Digital tools enable real-time collaboration and dissemination of ideas, which speeds up innovation and implementation of sustainable practices.

Finally, policies supportive of sustainable technologies and resource management need to be enforced consistently across the globe. This requires strong governance and international cooperation. Transparent regulation and monitoring can ensure that nations adhere to their commitments, creating a level playing field that encourages sustainable development.

Engaging in community and global efforts isn't just an option—it's a necessity. The challenges we face are too vast and complex to tackle in isolation. By coming together, communities and nations can leverage their strengths, share their successes, and work towards a sustainable future. This collective action is the bedrock upon which the future of technology and resource management must be built.

Conclusion

The journey through this book has been nothing short of enlightening. We've traversed the realms of critical materials, sustainable technologies, and the convergence of energy and environmental stewardship. All along the way, our goal has been clear: to illuminate the paths we must take to sustainably manage the vital resources that will shape our future.

In laying out the challenges and opportunities in chapters focused on resources like rare earth elements, lithium, and bio-based materials, we have highlighted the essential ingredients for tomorrow's technologies. The stories of innovation and the pressing need for recycling and recovery paint a vivid picture of a world where resource efficiency is not just an option but an imperative.

Solar and wind energy, the pillars of renewable power, showcase the incredible advancements in material sciences. From perovskite solar cells to composite wind turbine blades, the future of clean energy depends heavily on our ability to innovate in material composition and application. These examples underline that sustainability is not just about what resources we use, but how we use them.

Water and energy are intrinsically linked, and the chapters on the water-energy nexus remind us of the ongoing demand for clean water amidst a growing global population. Desalination, hydro, and tidal power innovations exemplify the cross-disciplinary approaches required to meet these demands sustainably. These technologies have the potential to quench the world's thirst and power our homes simultaneously, demonstrating how interconnected our resource challenges truly are.

Bioplastics and biofuels offer a glimpse into a future where our reliance on fossil fuels wanes. The rise of biodegradable materials highlights a paradigm shift towards products that work in harmony with our environment rather than against it. This shift is crucial for reducing the environmental footprint of our consumer-driven lifestyle.

The evolution of nuclear energy, especially in terms of safety and waste minimization, presents another critical frontier. With cleaner alternatives like thorium reactors on the horizon, nuclear energy could see a renaissance as a sustainable powerhouse. This transformation hinges on continued advancements in materials that ensure reactor safety and efficiency, positioning nuclear as a viable option in our diversified energy portfolio.

Fusion energy, often described as the holy grail of clean power, is tantalizingly close yet still elusive. The challenges in achieving fusion are monumental, but the potential rewards are equally vast. Innovations in superconductors and other essential materials drive this dream forward, representing humanity's relentless pursuit of the ultimate energy source.

The circular economy and resource efficiency chapter bring us back to the essence of sustainable living: reducing waste, increasing product lifespans, and rethinking the design of everyday items. By embracing principles of circular economies, we can create a loop where materials are continuously reused, dramatically decreasing our environmental impact.

Throughout this book, we've seen that mining isn't just about extraction; it's about innovation. Green mining technologies and urban mining initiatives show that even traditionally destructive industries can evolve into allies in our quest for sustainability. The ability to repurpose and reuse materials extracted from our urban environments may redefine the future of resource acquisition.

Alternative materials like graphene and nanomaterials push the boundaries of what's possible. Their remarkable properties promise to revolutionize multiple industries, highlighting the need for continued research and development. These materials represent the cutting edge of technological advancement and offer new pathways to solve age-old problems.

The concluding chapters explore the human element, policy, investment, and the digital revolution. The role of policy and governance cannot be overstated; it is the bedrock upon which sustainable practices are built. Meanwhile, the infusion of AI and big data into resource management heralds a new age of optimization and efficiency, making sustainable practices more feasible and widespread.

The interconnectedness of global resources means that local decisions have far-reaching implications. Ensuring equitable access to these resources, overcoming technical and economic barriers, and fostering societal adaptions are all critical steps toward a sustainable future. This book has aimed to outline these steps, providing a roadmap for action.

Case studies illustrate that innovation and sustainability are not merely theoretical. They are actionable, tangible goals being realized around the world. By learning from these successes and the lessons they offer, we can forge a path forward that is both innovative and sustainable.

In the end, envisioning a sustainable world is not enough. We must act. The citizen's guide encourages everyday actions and community engagement, showing that everyone has a role to play. By committing to these values, we collectively move towards a greener, more sustainable future.

As we close this exploration, let's be motivated by the possibilities and inspired to take meaningful action. The future of technology and sustainability is in our hands, and together, we can build a world where innovation and environmental consciousness go hand in hand.

Appendix A:
Appendix

In this appendix, we provide a comprehensive collection of supplementary materials and additional insights that complement and enhance the discussions presented throughout this book. It's designed to serve as a resource for those who wish to delve deeper into the intricacies of technology and sustainability, offering both a summary and expansion of key concepts introduced in the preceding chapters.

Table Summaries and Additional Data

Throughout the book, we've included various tables and charts to illustrate data points and trends. In this section, you'll find summarized versions of these tables alongside additional datasets that weren't included in the main text due to space constraints. This supplementary data can offer a clearer picture and further substantiate the arguments made.

Extended Case Studies

We touched on numerous case studies in Chapter 23. Here, we offer extended versions of these case studies, providing a more detailed look at the innovations and sustainability practices that are making a difference around the world. These extended narratives aim to provide a fuller understanding of the strategies and outcomes achieved, illustrating both successes and ongoing challenges.

Technical Specifications

The book covers a wide range of materials and technologies, many of which involve complex technical specifications. This section contains detailed descriptions of the technical aspects, including the chemical compositions, physical properties, and manufacturing processes. This is particularly useful for those with a technical background seeking to understand the fine details of the innovations discussed.

Further Reading and Resources

While the main text includes a Glossary and a list of recommended resources, this section expands upon those lists. You'll find additional books, articles, and websites that offer more in-depth information about specific topics covered in the book. Whether you're an educator, policymaker, or simply an interested reader, these resources are invaluable for further exploration.

Acknowledgments and Contributions

This work wouldn't have been possible without the invaluable contributions of a diverse group of experts, researchers, and practitioners. In this section, we acknowledge everyone who has provided their expertise, feedback, and support throughout the writing process. It's a testament to the collaborative effort needed to tackle the complex issues of technology and sustainability.

As you navigate this appendix, remember that the journey toward a sustainable future is both a personal and collective endeavor. Every piece of information here aims to equip you with the knowledge and inspiration to take meaningful action. By understanding the underlying science and the global context, we can all contribute to shaping a more sustainable and innovative world.

Finally, if you come across any terms or concepts that are unfamiliar, be sure to consult the Glossary section for clear definitions. And don't hesitate to explore the recommended readings for a broader view of the topics you find most compelling.

Glossary of Terms

Welcome to the "Glossary of Terms," a curated list intended to provide you with a clear understanding of the key concepts, materials, and technologies that are pivotal to the future of technology and sustainability. As we delve into the multifaceted world of resource management and sustainable innovation, this glossary serves as a quick reference to help you navigate through the complex topics covered in this book. Let's break down the terminology that will shape tomorrow's technologies and our planet's future.

Bioplastics

Bioplastics are a type of plastic derived from renewable biological sources such as plants or microorganisms. Unlike conventional plastics, which are made from petroleum, bioplastics aim to be more environmentally friendly and biodegradable.

Circular Economy

A circular economy is a sustainable economic system that prioritizes the recirculation of resources over the linear 'take-make-waste' approach. By designing products for durability, reusability, and recycling, a circular economy helps reduce waste and conserve natural resources.

Desalination

Desalination is the process of removing salts and minerals from seawater to produce fresh, drinkable water. Innovations in desalination technologies are crucial for addressing water scarcity worldwide.

Fusion Energy

Fusion energy is a form of power generated by fusing atomic nuclei, a process that occurs naturally in the sun. Achieving controlled fusion on Earth could provide an almost limitless and clean energy source, though it remains an elusive goal due to significant material and technical challenges.

Graphene

Graphene is a single layer of carbon atoms arranged in a two-dimensional honeycomb lattice. Known for its extraordinary electrical, mechanical, and thermal properties, graphene is often termed the 'miracle material' with vast potential applications in electronics, energy storage, and more.

Lithium

Lithium is a lightweight metal critical for the production of rechargeable batteries. It's pivotal in the battery revolution, powering everything from electric vehicles to portable electronics. Understanding its journey from extraction to application, as well as its environmental impacts, is essential for sustainable electrification.

Nanomaterials

Nanomaterials are materials engineered at the nanoscale, typically less than 100 nanometers. These materials exhibit unique physical, chemical, and biological properties due to their size and surface area, making them highly valuable in applications ranging from medicine to environmental cleanup.

Perovskites

Perovskites are a group of materials with a distinct crystal structure, widely explored for their potential in next-generation solar panels. They offer a promising alternative to traditional silicon, with the potential for higher efficiency and lower production costs.

Rare Earth Elements (REEs)

Rare Earth Elements are a set of seventeen chemical elements in the periodic table, essential for advanced technologies like smartphones, wind turbines, and electric vehicles. Despite their name, they are relatively abundant but challenging to extract economically and sustainably.

Superconductors

Superconductors are materials that can conduct electricity without resistance when cooled to very low temperatures. Their role in fusion reactors is vital, as they can generate extremely strong magnetic fields necessary for plasma confinement.

Turbine

A turbine is a device that converts kinetic energy from a fluid (like wind or water) into mechanical energy. In the context of renewable energy, turbines are crucial for generating electricity in wind, hydro, and tidal power systems.

This glossary is by no means exhaustive, but it includes some of the most pivotal terms you'll encounter as we explore the vast terrain of future technology and sustainability. Use it as a companion guide to deepen your understanding and inspire actionable insights.

Recommended Resources for Further Reading

Diving deeper into the landscape of future technology and sustainability requires more than cursory knowledge—it demands a comprehensive understanding and continual learning. This section provides curated resources to expand your horizon beyond the terminologies and concepts covered in the Glossary of Terms. Whether a scholar, engineer, policymaker, or an enthusiastic learner, these recommended readings will propel your journey towards sustainable innovation.

1. **"Materials for a Sustainable Future"** - This book is an essential starting point for understanding the critical materials at the heart of modern technological advancements. It explores the life cycles of these materials, environmental impacts, and innovations in recycling and recovery, making it a staple for anyone invested in technology and sustainability.

2. **"The Elements of Power"** by David S. Abraham - Focusing on the rare and valuable elements that drive our everyday technologies, Abraham's work emphasizes the geopolitical and economic factors influencing the availability of these resources. It aligns well with the earlier sections on rare earth elements and their importance to technological progress.

3. **"The Battery Revolution"** - This text outlines the journey of lithium, from mining to its crucial role in electrification and energy storage. It offers extensive insights into environmental implications and the future of battery technology, perfectly complementing the discussions in Chapter 2.

4. **"Sustainable Energy - Without The Hot Air"** by David J.C. MacKay - MacKay's pragmatic approach to sustainable energy solutions focuses on real-life applications and innovations in solar and wind energy. It serves as a practical guide for anyone looking to understand and implement renewable energy technologies.

5. **"The Water-Energy Nexus: Comprehensive Resources to Address Resource Scarcity"** - Bridging the gap between water and energy, this book explores advancements in hydro and tidal power, desalination technologies, and the potential of water as a sustainable energy source.

6. **"The New Plastics Economy"** by The Ellen MacArthur Foundation - As global concern for plastic waste rises, this book dives deep into the principles and practices of creating a circular economy for plastics. It is an invaluable resource for anyone interested in bioplastics and sustainable plastic management.

7. **"Thorium: Energy Cheaper than Coal"** by Robert Hargraves - Thorium reactors are often mentioned as cleaner alternatives to traditional nuclear energy. Hargraves addresses technological, environmental, and economic benefits of thorium-based energy, making it a crucial read for nuclear energy enthusiasts.

8. **"The Quest for Fusion Energy"** - This collection of research discusses the material challenges in achieving fusion energy, highlighting the vital role of superconductors. It offers a detailed analysis of the visionary approaches driving the race toward fusion reactors.

9. **"Cradle to Cradle: Remaking the Way We Make Things"** by Michael Braungart and William McDonough - This seminal work on circular economy principles encourages readers to rethink product design and lifecycle. It provides foundational knowledge for those

interested in innovative recycling technologies and sustainable resource management.

10. **"Urban Mining: A New Resource Paradigm"** - Focusing on extracting valuable materials from electronic waste, this book sheds light on urban mining as a sustainable alternative to traditional mining practices. It emphasizes methodologies and technologies pertinent to Chapter 10.

11. **"The Graphene Revolution"** - Delving into the potential of graphene, this book covers its applications in electronics, energy storage, and materials science. It is a perfect complement to Chapter 11, exploring the vast opportunities this miracle material offers.

12. **"Water 4.0: The Past, Present, and Future of the World's Most Vital Resource"** by David Sedlak - A comprehensive guide through the advancements in water purification alongside conservation techniques, crucial for sustaining our most vital resource, water.

13. **"The Right to Be Cold"** by Sheila Watt-Cloutier - A gripping narrative intertwining environmental issues with human rights, particularly focusing on the Arctic. This book serves as both an educational tool and a call to action regarding climate change and air quality.

14. **"Soil Not Oil"** by Vandana Shiva - Bringing attention to the interconnectedness of soil health, climate change, and sustainable agriculture, Shiva's work aligns with the efforts discussed in Chapter 14.

15. **"Investing in a Sustainable Future"** - Targeted at investors and policymakers, this financial guide explores the economics of sustainable technologies, funding strategies, and the impact of green investments, closely related to Chapter 17.

These resources, while diverse in their scope, converge on a single goal: fostering a sustainable, innovative future. As you delve into these readings, you'll gain a deeper understanding and be inspired to act towards managing the precious raw materials shaping our world. Each book and article serves as a stepping stone, offering both knowledge and motivation to partake in this crucial journey of sustainability and technological advancement.

www.ingramcontent.com/pod-product-compliance
Lightning Source LLC
Chambersburg PA
CBHW022250290526
45785CB00015B/498

9 781456 651701